PRINCIPLES OF

Applied Civil Engineering Design

Other Titles of Interest

How to Work Effectively with Consulting Engineers: Getting the Best Project at the Right Price (ASCE Manuals and Reports on Engineering Practice No. 45, updated edition). 2003. ISBN: 0-7844-0637-5. Price: $29.00.

Minimum Design Loads for Buildings and Other Structures, SEI/ASCE 7-02. 2003. ISBN: 0-7844-0624-3. Price: $98.00.

Ten Commandments of Better Contracting: A Practical Guide to Adding Value to an Enterprise through More Effective SMART Contracting. 2003. ISBN: 0-7844-0653-7. Price: $79.00.

Standard Guidelines for the Collection and Depiction of Existing Subsurface Utility Data, CI/ASCE 38-02. 2003. ISBN: 0-7844-0645-6. Price: $44.00.

Quality in the Constructed Project, Second Edition (ASCE Manuals and Reports on Engineering Practice No. 73). 2000. ISBN: 0-7844-0506-9. Price: $40.00.

Surety Bonds for Construction Contracts. 2000. ISBN: 0-7844-0426-7. Price: $49.00.

EJCDC Contract Documents
Standard contracts developed and periodically updated by the EJCDC represent the latest and best thinking of practicing engineers and attorneys on the subject of contractual relations between the parties involved in a project. http://ascestore.aip.org/

PRINCIPLES OF

Applied Civil Engineering Design

Ying-Kit Choi

Library of Congress Cataloging-in-Publication Data

Choi, Ying-Kit.
    Principles of applied civil engineering design/by Ying-Kit Choi.
      p. cm.
    Includes bibliographical references and index.
    ISBN 0-7844-0712-6
      1. Buildings—Specifications.   2. Construction contracts.   I. Title.

TH425.C44 2004
  624—dc22

                    2003063835

Published by American Society of Civil Engineers
1801 Alexander Bell Drive
Reston, Virginia 20191
www.pubs.asce.org

CONTENTS

v

FIGURES

TABLES

PREFACE

In 1984, I was hired by a nationally recognized civil/geotechnical consulting firm in Massachusetts. Three engineering degrees, all in civil engineering, and 2 years of teaching civil engineering at one of the best civil engineering universities in this country convinced me that I was ready for any assignment. My first task was to perform an engineer's cost estimate for an excavation to construct a new subway station in Boston. That provided the first indication that I was ill-prepared for the commercial consulting world. To complete this assignment, I had to estimate unit prices for dewatering, braced excavation, cofferdam protection, and miscellaneous earthwork items, and I had to estimate quantities based on the plan layout of the design. I had never heard of R. S. Means, whose construction cost data would be the basis for the unit price estimate. I was not familiar with the so-called bid schedule, which is the basis on which a contractor submits a bid and is paid for his or her work. Needless to say, that was quite an eye-opening experience for me, and, after asking many questions and making many mistakes, I completed the assignment in excess of the allowed budget and beyond the assigned time.

After many small assignments in traditional foundation investigation projects in that first year, I found myself as a project engineer for a fast-track dam rehabilitation project located in Virginia. The position required me to prepare construction plans and specifications in fewer than 3 months. Before that assignment, I had never prepared construction drawings, nor had I ever written any technical specifications. There was a lot of quick learning on my own during this madly paced assignment. I quickly discovered that the only resources available to me were the more-experienced designers in the company and whatever examples of similar projects I could find in other project files. Ironically, even though life during this design assignment could be described as extremely unpleasant, I soon discovered near its end that I actually enjoyed sitting behind a drafting table creating construction drawings. The feedback that I received from management at the end of that assignment was that I should be more efficient in doing design work.

That was the beginning of a long tenure of learning experience in civil engineering designs for me. During that tenure, my emphasis was in civil/geotechnical design and construction engineering. As I developed into a senior designer, I discovered that mentoring junior staff designers and working with computer-aided drafting (CAD) drafters would have been more efficient if there had been a design reference that I could have used as a teaching tool. The dream of writing a book on applied civil engineering design developed into reality when I decided to be self-employed, without the day-to-day responsibilities of project management, marketing, and proposal writing typical of most senior professionals at such a point in their careers. When the book proposal and manuscript were submitted to American Society of Civil Engineers (ASCE) for review, the feedback from all of the reviewers was overwhelmingly favorable and supportive, demonstrating the need for such a reference in the civil engineering design profession.

The primary target audience for this book is young civil engineers and civil engineering students who want to learn how to prepare final design documents. My ultimate hope is that applied civil engineering design can be taught in a civil engineering curriculum, so that young professionals will not have to learn on the job. This book is a teaching tool, and I firmly believe that abstract concepts and principles should be taught with examples and illustrations, which are plentiful in this book. Most of the examples and illustrations used in this book draw heavily from my own design experience and projects. While most of the design principles represent standard and conventional practice, there are also many design philosophy and design approaches that are my opinion. Although the philosophy and approaches are merely one man's opinion, they have worked well for me in my design career.

Besides young engineers, this book will benefit those involved with the design process—namely, the more senior design reviewers, drafters, cost estimators, and specification writers. Civil engineering design requires team work, and each team member has a unique role and set of responsibilities. I attempt to define the roles and responsibilities of separate design team members so that each will perform within his or her assignment. Throughout my design career, I was appalled that some design projects were not always staffed appropriately, and the results were usually cost overruns, delays, construction problems, and claims. I believe that some of these problems are caused by management's lack of understanding of the design process. With a better understanding of minimum qualifications and clear definitions of roles and responsibilities, I wish to educate the managers and decision-makers as well.

This book also will be valuable to contractors, particularly for their young project managers and project site engineers, many of whom are new graduates and are inexperienced in reading and interpreting construction drawings and technical specifications. Like young civil designers, these contractor personnel will have to learn on the job, with a steep learning curve. Although experience learned on the job is an essential part of one's development into a good construction manager, this book provides the developing site engineer a valuable insight into the basic principles from a designer's point of view. It also provides a background for them to effectively communicate with the designer during construction, prepare record drawings, prepare change orders and submittals, and estimate construction costs and quantities.

This book may also be helpful to owners of civil engineering projects. Whereas it is the responsibility of the design engineer to provide all necessary technical services from the inception of a project to its completion, the owner still plays a significant role. His or her responsibilities include: funding and financing design and construction; applying for necessary permits and interacting with regulatory agencies; managing the performance of the engineer; establishing project design criteria and requirements; participating in the development of the construction bid schedule for measurement and payment; and managing the financial aspect and performance of the contractor during construction, including progress payments, change orders, and claims.

The vital interaction of a project owner, engineer, and contractor makes it necessary for an owner to understand the key decisions and recommendations provided by his or her engineer and the construction issues affecting the cost of the project. Of particular interest to the owners will be the following topics: adequate funding of characterization of a project site and the construction cost implications of an inadequately characterized site; effective scheduling of the engineering design and preparation of the plans and specifications to allow the engineer adequate time to prepare a complete set of documents for bidding; cost implications of fair and risk-sharing approaches in contract specifications; and the strategy of bid schedule items preparation to minimize potential claims during construction.

This book is organized into four parts. Part 1 discusses the need for and scope of the book, the data that are needed for design of a civil engineering project, and how the construction drawings, specifications, and cost estimate fit into the overall scheme of a set of bid documents. Part 2 deals with the details and mechanics to prepare a set of construction drawings for a civil design project. Drawing production techniques are introduced and illustrated with examples. The use of computers and CAD is discussed. Part 3 deals with the preparation of technical specifications, with emphasis on using the Construction Specification Institute (CSI) format. Bid schedule and measurement and payment provisions are particularly emphasized. Part 4 deals with preparation of an engineer's cost estimate, including estimating quantities and developing unit and lump-sum prices. The use of various allowances and contingencies is also discussed for different levels of the design. In each of Parts 2, 3, and 4, the discussions of the interrelations among drawings, specifications, and cost estimates illustrate that these three documents and processes must be part of a coherent and coordinated set of documents intended to effect the successful construction of a civil engineering project.

ACKNOWLEDGMENTS

The author is deeply grateful to the following friends and professional colleagues who contributed to the review of this book. All of these individuals were practicing professionals with busy schedules, and yet they graciously provided the critical review on the manuscript prior to submission to ASCE. Mr. Alton P. Davis, Jr., whom the author considers his mentor, reviewed the entire draft manuscript, and provided many valuable comments and suggestions. Mr. Gregg Batchelder-Adams, a former colleague and an excellent civil/geotechnical designer, reviewed Parts 1 and 2 of the draft manuscript. Mr. Ken White, a construction manager and former contractor, reviewed Part 3 of the draft manuscript. Mr. Roy Watts, a former contractor and a professional cost estimator, reviewed Part 4 of the draft manuscript. Mr. Michael Boulter, one of the most talented CAD drafters the author has known, expertly prepared all of the figures. The assistance of these professionals is graciously acknowledged.

The preparation and production of the manuscript would not have been possible without the emotional support of the author's wife LeEtta. Her constant encouragement, love, and understanding are affectionately acknowledged.

PART **1**

INTRODUCTION

CHAPTER 1

PURPOSE AND SCOPE

1.1 Applied Civil Engineering Design

Applied civil engineering design is a multidisciplinary process involving detailed analysis, judgment, and experience aimed at producing construction drawings, technical specifications, and bid schedules required to allow contractors to bid and construct physical projects. Civil engineering encompasses such disciplines as structural engineering, geotechnical engineering, water resources engineering, environmental engineering, transportation engineering, and many related subspecialties. The emphasis of this book is on heavy civil engineering construction projects. There is no precise definition for *heavy construction*. Ringwald (1993) listed several characteristics of heavy construction projects, as follows:

- Equipment cost, expressed as a percentage of total project cost, is about 10 times higher in heavy construction than in building construction.

- Heavy construction projects tend to spread out in a horizontal plane, as compared with the vertical nature of a building.

- Heavy construction is usually performed for a public owner, whereas building work is usually performed for a private owner.

- Heavy construction documents are prepared by engineers, whereas building documents are prepared by architects.

- Heavy construction is much more weather-sensitive, and allows far fewer working days per construction season than building construction.

The characteristics listed above provide an applicable description of the types of construction projects intended for this book. Heavy civil engineering construction projects include roads and highways, dams, foundation excavations for buildings, water and wastewater treatment plants, tunnels, bridges, airports, power plants, pipelines, drainage and flood control facilities, and urban development. A *civil engineering designer* is a specialized and experienced professional engineer capable of producing construction documents for these projects.

University civil engineering curricula, even at the graduate level, do not currently provide the necessary training or skills for a civil engineer to practice design immediately upon graduation from a bachelor's or master's degree program. Frequently, the design skills of a civil designer are gained through many years of design practice and mentoring under a senior professional, combined with experiences gained through field construction observations. Certainly, the basic technical background and courses in engineering graphics, surveying, engineering contracts and specifications, and mathematics that one learns in a typical engineering curriculum are important in building these design skills. However, contrary to other design disciplines, such as mechanical, electrical, or architectural design, the production of civil engineering design documents is not taught in academia, nor are there readily available guidelines for young practicing civil engineers to gain these vital skills. Most universities offer no practical/applied civil engineering design courses, and no references on this critical subject are currently available.

1.2   Objectives

There are three main objectives in this book:

1. Provide beginning and practicing civil engineering design professionals with a reference that includes definitions of basic principles and can be used to prepare construction drawings, technical specifications, bid schedules, and engineer's construction cost estimates with consistent style and format.

2. Provide recommended guidelines and design approaches in preparation of quality construction documents.

3. Provide a starting point in a civil engineering curriculum on which an engineering course on applied civil engineering design can be based.

It is important to point out that this book should be considered as a pioneering publication on a subject that has been practiced by the profession for many years. There are numerous books that touch on portions of the subject matter, but none tie all the parts together. For example, there are many books written on engineering graphics, but the emphasis of those books is on mechanical or architectural drawing, not civil engineering layout drawing. There are many books written on engineering contracts and specifications, but typically, the emphasis is on contracts and administration-related matters, not technical specifications. There are many books written on engineering surveys, but not the application of survey techniques to civil design and project layout control. Finally, there are many books written on estimating quantities or costing work, but a compilation of the entire cost-estimating process (establishing a bid schedule, estimating unit price and lump-sum prices, quantity takeoffs, and writing measurement and payment clauses) currently does not exist specifically for civil engineering projects.

1.3   Special Features and Approaches

The unique features of this book are

*   the application of traditional engineering graphics to civil design layouts;

*   the consolidation of three main components of final design—construction drawings, technical specifications, and engineering cost estimating—under one cover

and into a coordinated and interrelated set of documents that can serve as a starting reference for young professionals;

- the use of heavy earthwork projects, such as dams, drainage and flood control projects, and highway projects, as examples and illustrations;

- the discussion of technical specification writing using the standard Construction Specifications Institute (CSI) format; and

- the inclusion of teaching materials that can be used to form the basis for a one-semester course for senior civil engineering students or a graduate course in civil engineering applied design practices.

Because this book is also intended to be a teaching tool, numerous examples and figures are used to illustrate key points. General rules and guidelines are explained using specific examples. Overall, the author draws on his own practical experience, design philosophy, and approaches to produce a set of quality construction documents that are coherent, well-coordinated, easily understood by the contractor, and contractually fair to all three concerned parties (the owner, the engineer, and the contractor) in a construction project.

1.4   Use of Design Guidelines

This book introduces the conventional methods, styles, and formats for producing construction drawings and technical specifications. None of them is new, and many of them are not standardized. Many design firms and agencies have their own drafting standards and specification formats and styles that they've developed through many years of use and refinement. It is not expected that these entities will change their standards or formats for the guidelines and recommendations given in this book. That is not the intention of this book. Nor is it the intention of this book to standardize drafting of civil drawings (e.g., line weights, lettering style and heights, hatchings and symbols). Rather, design guidelines and recommendations on the preparation of drawings and specifications are introduced in this book for the following reasons:

- To show students and young engineers some of the available tools and methods used to produce these documents. Young designers can use them as a starting point in their work, or they can use them to understand specific standards that they need to follow in their own firm or agency.

- To show readers that are not engaged in design (e.g., young contractors, cost estimators, owners) how to understand and interpret construction documents.

- To show computer-aided drafting (CAD) drafters of other disciplines (e.g., architectural, mechanical, structural) the basic information and styles typically needed for developing civil drawings.

- To use them as a starting point for dialog, future improvements, and possibly some form of standardization. The CSI already has taken an important step in standardizing the preparation of specifications and architectural drawings, and similar effort should be extended in applied civil engineering design.

For the practitioners of civil design, particularly those that are directly involved with production of the drawings and specifications, readers are expected to use their own judgment and experience to decide what is and is not acceptable. To a student who is learning how to draw a plan or cross section, whether by hand or on the computer, it is important that he or she starts out with some basic techniques and fundamental skills that can be used as stepping stones for his or her future professional development.

CHAPTER 2

CONTRACT DOCUMENTS

2.1 Purpose

Construction contract documents are used for bidding and construction of projects. During bidding of the procurement phase, the contract documents are part of the bidding documents; after the contract is awarded, the bid documents become the contract documents. The administrative components of the contract documents are the general conditions and supplemental conditions, and the technical components are the construction drawings and specifications. The technical requirements are prepared by the engineer during final design. Preparation of construction drawings and specifications are the primary features of Parts 2 and 3, respectively, of this book. These two components of the contract documents are interrelated, and it is important to understand the general relation between the two before proceeding with preparation of the technical documents. For this reason, an overview and background on contract documents are given in this chapter; however, an in-depth treatment of the administrative documents (so-called boiler plate documents) is beyond the scope of this book.

An understanding of the process of competitive bidding is also important in learning how to prepare a bid schedule and the related measurement and payment provisions. During final design, the engineer also prepares payment arrangements for construction projects. Preparation of a schedule for bidding, measurement and payment provisions, quantities, and estimated construction costs are the primary features of Part 4 of this book.

The process considered in this chapter is for the conventional design-bid-build construction procurement arrangement that has been used successfully for many years. Although some of the bid documents and bid processes are still applicable for the design-build method, no further reference will be made on the subject of design-build in this book.

2.2 Competitive Bidding

The purpose of the competitive bidding process is to select a contractor to construct the project. Typically, the owner, or his or her representatives or agents, manages the bidding process. Briefly, the competitive bidding process consists of the following steps:

- **Advertising** A construction project is advertised to solicit interest or bids from interested contractors. The advertisement generally contains information on the owner, project location, scope of work, minimum qualifications, and a range of construction costs. The cost range indicates the size of the project, which provides a means for the bidder to determine whether he or she can be bonded to perform the work. The cost range also prevents a small contractor from bidding on a large project. The advertisement also indicates where bid documents (see Section 2.3) are made available to prime contractors, subcontractors, material suppliers, manufacturers, and distributors.

- **Prebid meeting** A meeting with prospective bidders is conducted at the beginning of the bidding period. The purpose of this meeting is to explain the work scope and other administrative requirements to the bidders; it usually includes a visit to the project site. Some owners make this meeting mandatory, but others, such as the federal government, do not. Questions raised by bidders during this meeting, and the answers to those questions, are generally recorded and distributed to everyone on the plan holder's list.

- **Issuance of bid amendments** Additional information provided to the bidders during bidding are called *amendments* or *addenda*, and are considered legal parts of the bid documents. Amendments may include answers to bidders' questions, design changes, changes in bidding period or construction period, new field information, or other added requirements (e.g., permits). To allow bidders adequate time to respond and adjust their bids where appropriate, it is generally a fair practice to extend the bidding period when a bid amendment is issued close to the bid opening deadline. It is important that a record is made of all those receiving the addenda, either through the use of certified mail or records of fax transmission.

- **Bid opening** Bid opening occurs at, or shortly after, the bid submittal deadline. Bid opening can be open or closed to the public, depending on the owner of the project. In general, all bids received are summarized in a form called a *bid tab* using the format of the bid schedule (see Chapter 25). At this time, the bidder that submits the lowest total bid is declared the "apparent low bidder." It should be noted that any changes in the design and any changes to the work requirements after the bid opening are considered "changes," and should be properly handled through "change orders."

- **Bid evaluation** Owners, with assistance from their engineers, generally evaluate the low bid and other bids to determine: whether the apparent low bidder has adequately responded to the work requirements in his or her bid; whether any unbalanced bidding (see Chapter 25) has taken place; or whether there are other omissions of bid submittal information. When it is necessary to obtain the basis for, and confirmation of, any items in question, the owner requests a bid verification from the apparent low bidder. In some cases, the owner will meet with the apparent low bidder to better understand the bid and negotiate changes prior to award.

- **Contractor selection** If the owner is satisfied with the bid, the apparent low bidder is selected as the contractor. A construction contract is established between the low bidder and the owner. For the engineer, the bidding process ends here, and construction management begins. Construction management will not be discussed in this book.

2.3 Bid Documents

Bid documents are information furnished to the bidders during the bidding period. These documents typically include an Invitation to Bid, Instructions to Bidders, Bid Forms, general conditions, supplemental conditions, construction drawings, technical specifications, and amendments. All contractual, administrative, and technical requirements are contained in these documents. These documents define the duties and responsibilities of all parties involved. For federal construction, the contractual documents are contained in the construction contract clauses of the Federal Acquisition Regulations (FAR), but they contain similar types of information as the documents for private-sector construction. The following are brief descriptions of these documents. Detailed descriptions are available in many publications for construction management (Fisk 1992).

Invitation to Bid

This document is used in advertising to solicit bidders for construction work (see Section 2.2).

Instructions to Bidders

This document contains procedures for bidding, basis and criteria that will be used for bid evaluation, and guidance to bidders for other relevant information contained in the bid documents.

Bid Forms

This document is used by the bidders to submit their bid prices and to document the basis contained in their bids. In many cases, a *bid schedule* is attached as part of the bid forms. The bid schedule (see Section 19.2) contains all of the payment items, quantities estimated by the engineers, measurement units, and owner-defined allowances for specific items.

General Conditions and Supplemental Conditions

The general conditions and supplemental conditions are provided to bidders during bidding because they will become part of the legal contract documents after award of the contract (see Section 2.4).

Drawings and Technical Specifications

Drawings and technical specifications contain all of the technical requirements to construct the project. It should be noted that the technical specification package should also include reference data collected during design and investigations (see Chapter 20).

Amendments

After the contract is awarded, the bid amendments (see Section 2.2) become part of the contract documents.

2.4 Contract Documents

An engineering construction contract typically contains the following documents: agreement, general conditions, supplemental conditions, amendments, bid schedule, construction drawings, and technical specifications. All of these documents are considered legal documents, and the contents of these documents should be carefully compiled (or conformed) to avoid disputes, ambiguities, conflicts, and unnecessary information. Generally, the agreement, general conditions, and supplemental conditions are *contract forms* furnished by the owner, and the bid schedule, construction drawings, and technical specifications are prepared by the engineer. Together, they define how construction work will be performed and completed, how the contractor will be paid, the project schedule, bonding and insurance requirements, construction management and inspection, claims procedure and timing, and actions for breach of contract.

The agreement, signed by the owner and the contractor, is the legal document that takes precedence over all other documents in a construction contract. It makes references to all other pertinent documents, and is usually a standard form that varies for different owners.

The general conditions of an engineering construction contract define the duties and responsibilities of the owner, contractor, and engineer. They address all issues related to the administration and management of the contract, and include such items as bond and insurance; procedures for changes in work scope, schedule, and prices; warranty and guarantee; payment procedure and method; and dispute resolution. It is not the intention of this book to examine and scrutinize the contents of the general conditions, except in instances in which they affect the drawings, specifications, and engineer's cost estimate. Like the agreement, the general conditions are contained in standard documents that normally do not significantly change from project to project for a particular owner.

The supplemental conditions (also called *special provisions*) should be considered an extension of the general conditions, and are used to address site-specific requirements and unique characteristics for each construction project. In other words, any deviations from the general conditions should be handled in the supplemental conditions. Examples of items contained in supplemental conditions include specific bonding and insurance requirements, liquidated damages, project permits, local laws and regulations, site restrictions, coordination with other work on site, and site safety.

It is important for the engineer to understand all of the requirements in the general conditions and supplemental conditions so that they are consistent with the construction drawings and specifications. To avoid costly changes caused by conflicts and inconsistencies among these documents, coordination of the contract forms and the technical documents should be done before bidding.

For the resolution of conflicts and inconsistencies among various contract documents, the contract usually contains a definition of the hierarchy of these documents. With some exceptions, the typical hierarchy of contract documents, in order of decreasing precedence, is (1) agreement, (2) general conditions, (3) supplemental conditions, (4) amendments, (5) technical specifications, and (6) construction drawings.

2.5 Engineers Joint Contract Documents Committee Documents

There are numerous variations of bid documents and contract forms among different government agencies and among private-sector owners. In fact, most engineering firms have their own versions of the documents that they use for their clients. Because these documents are legal documents, any problems that arise out of a dispute for conditions in these documents may be subject to interpretation in a court of law. In an attempt to provide standard construction documents to the engineering industry, a group known as the Engineers Joint Contract Documents Committee (EJCDC) was organized. The EJCDC consists of National Society of Professional Engineers, American Council of Engineering Companies (formerly

American Consulting Engineers Council), American Society of Civil Engineers, Construction Specifications Institute, and Associated General Contractors of America.

EJCDC documents include bid forms, instructions to bidders, general conditions, and agreements between the owner and contractor. Some of these publications are used as guidelines, and some can be used in contract documents without changes. All of these documents are the products of many years of study, research, legal interpretations, and court decisions, and are endorsed by most engineering professions and contractors. The contractors' endorsements suggest that these documents represent not only the interests of the owner and the engineering profession, but also the interests of the contracting industry as well.

EJCDC documents are available through the member organizations, such as the American Society of Civil Engineers Publications Department (see List of Resources).

2.6 Permits

Project permits are typically included as part of the contract documents (see Section 2.4). In most civil engineering construction, project permits must be obtained by the owner prior to beginning construction. Most of these permits are issued by regulatory agencies for environmental control, and stipulations are included in the permits as conditions for construction. For example, the 404 Clean Water Permit issued by the U.S. Army Corps of Engineers is required if construction will take place in or near regulated features such as wetlands. Stipulations contained in the 404 Clean Water Permit may include limits on turbidity of construction water (e.g., dewatering discharge, runoff) discharged into streams, limits of wetlands that can be disturbed, and protection of aquatic and riparian vegetation and fisheries. These conditions must be enforced during construction, and the owner will usually pay fines associated with noncompliance. In some cases, failure to comply with these conditions will result in a construction work *stop order.*

Because of the consequences of a permit violation, all owner-acquired permits should be made part of the bid and contract documents so that appropriate actions and costs are included in contractors' bids. Owner-acquired permits that are added after bid opening are regarded as changed conditions, and the contractor is entitled to adjustments to the contract prices or time to abide by the requirements of the added permits.

It is important to point out that this discussion excludes the permits that the contractor is required to obtain during construction. Those permits are the sole responsibility of the contractor, and are required for hauling on, and access to and from, local and state highways; blasting; quarry development; disposal of dewatering and hazardous wastes; etc.

CHAPTER 3

CHARACTERIZATION OF PROJECT SITE

3.1 Site Characterization

This chapter describes the key characteristics of a civil engineering project site that are important for design. Central to characterizing a site for a civil design project is the definition of the existing conditions that are relevant to that particular project. Most of the effort in characterizing a site involves understanding what is in the ground and describing the topography of the ground surface (see Sections 3.5 and 3.6). There are two main reasons for adequately characterizing a site during design: to collect geologic and geotechnical data for engineering design and analysis, and to define reasonably the amount of construction work needed so that changes during construction will be minimized.

If a project site has not been used in the past, it is necessary to perform a geologic study (see Section 3.2) prior to performing a subsurface investigation (see Section 3.3). If a project site has been used previously, it is useful to perform research (see Section 3.4) on how the site was developed in the past.

To maximize the information obtained while conserving investigation costs, site characterization typically is accomplished in phases. Various levels of investigation are used during preliminary design, feasibility level design, and final design (see Section 3.7).

3.2 Geology

The importance of geology in civil engineering projects is well-established (Legget 1983). Because of the existing information that is readily available, most projects in developed urban settings do not require a geologic investigation, and a literature search may be all that is required. However, when a project is located in an undeveloped area, a geologic investigation is vital to the understanding of geotechnical and foundation issues that are relevant to the project. A geologic investigation may include geologic reconnaissance, a literature search, aerial photography, detailed structural geologic mapping, and trenching and laboratory testing. The level of investigation would depend on the size of the project, the complexity of the geology, and the design's dependence on geologic factors, such as fault activity estimates. In a seismically active region and for projects that impose large risks to public safety (e.g., dams and nuclear power plants) a seismotectonic assessment usually is required to determine the seismic design loading conditions.

Engineering geology is the discipline of applying knowledge of geology to engineering problems. In most cases, an understanding of the geologic setting for a particular project can identify potential problems. For example, the subsurface condition for a glacial till environment is highly variable, with significant amounts of cobbles and boulders that may impact the feasibility of some types of foundations. Young lacustrine deposits yield low strength and large settlements. Some windblown deposits are collapsible upon wetting and are highly erodible by flowing water. The existence of gypsum layers in rock formations may lead to the development of excessive foundation leakage and even dam failure. In the western United States, claystone and clay shale are prevalent, and these materials frequently give rise to foundation swelling problems and slope stability problems.

3.3   Subsurface Investigation

In a civil engineering project, subsurface investigation is used to obtain geotechnical information for foundations and earthwork design, to evaluate the constructability of excavations and backfill, and to estimate groundwater problems and mitigation methods. Subsurface investigation includes drilling and test pit excavation, field and laboratory testing, field monitoring of groundwater levels and quality, and geophysical investigation. The technical reasons for an adequate subsurface investigation and the methods of subsurface investigation (Hvorslev 1949; Clayton 1982; Winterkorn 1975; Terzaghi 1996; Head 1980) are not discussed in this book. Issues pertaining to constructability, construction problems, site safety, construction document preparation, and cost estimate are discussed herein.

After foundation analysis and the selection of the type of foundation, many design- and construction-related questions must be answered by a civil engineering designer:

- *How is the foundation built?* A geotechnical designer should be familiar with the various methods that are likely to be used for construction. A geotechnical design feature, no matter how attractive it is technically and economically, should not be used if the designer does not know how it is built. That does not mean that new construction technologies or innovations should be avoided. Rather, geotechnical designers should talk to specialty contractors, product manufacturers, or peers to obtain the information required to understand the processes and performances of these new methods or products.

- *Can a design from a previous project be reused?* A common error made by inexperienced designers using precedents is to ignore the special circumstances and conditions that are unique to a particular site. For example, a clean sand drainage blanket may be effective for seepage control of a silty and clayey foundation subgrade, but is not effective for a sandy foundation. Failure to investigate adequately a foundation's unique characteristics could result in a defective design with potentially damaging effects.

- *What excavation method is appropriate?* Many designers maintain that selection of an excavation method is not their concern because means and methods of the contractor are the responsibility of the contractor. To a certain extent, this is correct. Construction documents should be prepared to allow a contractor the freedom and flexibility to use whatever method and equipment he or she deems appropriate for a given site. This approach of not specifying the means and method of construction usually results in the most cost-effective construction method (see Section 15.5). However, when a designer or a cost estimator is required to estimate the cost of excavation, it becomes necessary to assume a specific method of excavation as the basis of the cost. The cost of excavation varies greatly for different types of soil materials, different soil densities or stiffnesses, and different degrees of saturation. Earth excavation methods, such as backhoe excavation, would be inappropriate for hard bedrock. Soft rock can be excavated by a hydraulic excavator or ripper, but hard rock requires blasting. Because of bearing pressure limitations, only certain construction equipment and excavation methods can be used for very soft subgrades. Without adequate subsurface investigation, the excavation cost cannot be estimated reliably, and the consequence is usually a construction claim for changed conditions or excessively high bids to account for unknown conditions.

- *Is the site safe for construction?* Site safety is a construction issue (see Section 15.12). It is discussed here in relation to design of earthwork. Most civil drawings depict a maximum slope angle for temporary excavation, and the contractor is also required to abide by the appropriate Occupational Safety and Health Administration (OSHA) regulations on temporary excavated slopes. A stable temporary excavated slope depends on many factors, including type of soils, location of groundwater, duration of exposure, and external loads. A designer should call out any special excavation requirements based on known ground conditions, such as bracings, dewatering, and restricting the depth of cuts, and there should be provisions in the contract documents to compensate the contractor for these requirements. Without adequate subsurface investigation, the contractor may not adequately anticipate safety requirements during bidding and performing the site work, increasing the likelihood of safety problems and claims during construction.

- *How is the site dewatered?* The issue of construction dewatering should be addressed during design, not during construction. Except under some special circumstances, selection of a dewatering scheme is the responsibility of the contractor. With the information obtained from subsurface investigation, a designer prepares the appropriate dewatering specifications and requirements, and the appropriate cost of dewatering is estimated accordingly. With that same information, a contractor also prepares his or her dewatering bid cost. For example, pumping from sump pits may be adequate for excavating into a clay foundation, but well points may be required for excavations into more pervious foundations below the groundwater table. It is almost impossible to bid on dewatering or to provide an engineer's cost estimate of dewatering with inadequate groundwater information and inadequate understanding of the subsurface materials.

- *How accurate are the excavation and backfill quantities?* An inadequately investigated foundation usually results in quantity overrun or underrun for excavation and backfill. For example, an unexpectedly shallow bedrock will require less earth excavation and more rock excavation, and an unacceptable subgrade will require overexcavation and replacement with structural backfill. Almost all contract documents have allowances for a contractor to renegotiate his or her unit prices for earthwork items that are significantly different than the bid quantities (see Section 22.5). Most of the time, the owner will end up paying higher unit prices for renegotiated work items.

So, from a civil design standpoint, what subsurface information is required to characterize a site? The following is a partial list of relevant subsurface data, in addition to requirements for engineering design of the structures:

- Subsurface conditions to the design limits of excavation. If bedrock is shallow and above the limits of excavation, it may not be necessary to explore all the way to that limit, depending on the types of rock.

- Laboratory tests to supplement the field descriptions of the various materials encountered in drilling or test pits, such as natural water contents in borrow pits.

- The effort of excavation should be characterized using quantifiable indices, such as the standard penetration blow counts from test borings. Soft foundation subgrade could require special excavation provisions, such as low-tire pressure equipment or support mats. Hard or stiff foundation subgrade may require a ripper or other special tools, and will affect the production rate of the excavation equipment. This information is particularly important for rock excavation. Soft rock, such as a weakly cemented sandstone or a weak claystone, may be excavated using a hydraulic hoe-ram or backhoe. Hard rock, such as a competent granite or limestone, requires blasting. Useful information for tunnel excavation includes coring rates, joint frequency and spacing, hardness, and unconfined compressive strength. A contractor bidding on a project involving rock excavation would need most or all of these data to estimate effectively the actual cost and method of excavation.

- Groundwater conditions at the limits of excavation. Groundwater conditions observed during test borings and test pit excavation should be recorded. Water levels monitored with observation wells are also useful to characterize the long-term stabilized conditions. Any changes in surface features, such as streams, irrigation canals, lakes, etc., should be noted in the monitoring data for any correlation of groundwater level with these features.

- Locations of buried utilities and buried structures.

- When the use of sheetpiles, foundation piles, or piers is anticipated, the ground conditions should be known to at least the bottom of the piles. The ability to drive piles into the ground without sustaining damages should be demonstrated in the design or provision made for a pile test program. Critical to the installation of driven piles is the presence of cobbles and boulders that should be made known in the documents. The presence of cobbles and boulders increases the chances of damaging piles during driving, introduces questions on what refusal criteria are considered adequate for acceptance, and greatly increases the potential for claims.

3.4   Prior Site Use Research

The reasons for researching the prior usage of a project site are threefold:

- To clear the site for any potential cultural or archaeological resources that are of value;

- To investigate the potential of environmental contamination and hazardous waste;

- To identify any buried structures that would affect the design and construction of the project.

In the United States, cultural and archaeological features may be related to Native Americans, early American settlers, or the Civil War. To most untrained eyes, the significance of most of these features is not obvious, and therefore, is often overlooked in a site survey or a field visit. For example, a circular pile of rocks may be an ancient fire ring used by Native Americans. Or, a trench in a hillside in an eastern state may be a civil war trench with great historical value. In an early planning phase of a civil engineering project, it is important to perform a screening-level study by archaeologists to determine whether the project site has any potential of archaeological or cultural significance. Most states now have an agency known as State Historical Preservation Office (SHPO) that provides guidelines and assistance, and this office should first be consulted at the beginning of the data collection phase of most civil engineering projects involving seemingly undeveloped land.

Investigation of site contamination requires field reconnaissance, sampling, and testing, and it requires knowledge of inorganic and organic chemistry, toxicology, and biology. These studies should be performed by specialists. A contaminated site, if discovered, should be adequately mitigated prior to redevelopment for a civil engineering project, and before authorizing the construction contractor to proceed.

Investigation for buried structures requires a search for previous construction records, drilling boreholes, and test pit excavations. Where it is not feasible to use boreholes or test pits, geophysical methods such as ground penetration radar (GPR) (Spangler 1982) may be used to identify buried objects. In many cases, GPR can be used to optimize other investigation programs.

3.5 Topographic Survey

Topographic survey is the applied engineering discipline to characterize the existing ground surface, to determine the configuration (relief) of the surface of the ground to document locations of field investigations, and to locate the natural and man-made features on the ground. The purpose of topographic survey is to obtain a map of existing conditions that is used as a starting base map for civil design projects (see Section 3.6).

The characterization of topographic relief defines the existing grade that determines the extent of cuts and fills required. Methods of topographic survey include transit-tape, transit-stadia, cross-sectioning, total station, global positioning system (GPS), and photogrammetry (ASCE 1999; Anderson 1997; Brinker 1969). With the advance of electronics, satellites, and lasers in recent years, topographic survey techniques have been developed into a very efficient and accurate operation, and now the most commonly used methods for topographic survey are total station, photogrammetry, and GPS. Topographic surveys for preparing construction documents should be performed by a licensed surveyor.

Depending on the intended use of the topographic maps, it might be necessary to survey all of the natural and artificial features on the ground. Natural features include vegetation, streams and drainage features, rock outcrops, and lakes. Surveying the limits and nature of vegetation, such as trees, shrubs, wetland, and marshland, is important for the land that will be developed or disturbed. The limits of trees and shrubs are used to define the limits of clearing and grubbing work for site development. With the enactment of the National Environmental Policy Act (NEPA), the preservation of natural resources, such as wetland and marshland, becomes an important consideration in the permitting, layout, and design of most civil engineering projects. Depending on the nature of the site development, the definition and delineation of wetlands may require the use of specialists (e.g., biologists and ecologists) who will delineate these limits before they are surveyed.

Areas that have been disturbed or developed contain artificial and cultural features that should be included in a topographic survey. These features may include roads, trails, buildings, utilities (e.g., power lines and telephone lines), bridges and highway overpasses, water conveyance structures (e.g., ditches and canals), stone walls, fences, and property markers. The locations of these features may be important in the early planning and design phases of a civil engineering project when determining conflicts, interference problems, site restrictions, permanent and temporary easement requirements, property ownership, and permit requirements.

The starting point for a topographic survey is the establishment of vertical and horizontal controls. Controls can be tied to national, state, or county datums and grid systems, or they can be local, depending on the project requirements. Regardless of whether local, state, or national datum are used, the basis on which the survey is performed should be clearly stated prior to field work. Horizontal controls are provided by two or more monuments with known coordinates; similarly, vertical controls are provided by two or more benchmarks with known elevations. Frequently, the same two monuments or benchmarks contain both coordinate and elevation information.

In the United States, standard elevations and horizontal coordinates are still expressed in feet. In civil engineering and surveying practice, horizontal coordinates are expressed in terms of northings and eastings. Northings are distances measured in the north-south direction that increase from south to north. Eastings are distances measured in the east-west direction that increase from west to east. A point in plan view is represented by a northing coordinate, easting coordinate, and an elevation, similar to the x-, y-, and z-coordinates in mathematics. Figure 3-1 shows an example of three points that are represented by this coordinate system.

The newest vertical control datum is the 1988 North American Vertical Datum, which is also referred to as the National Geodetic Vertical Datum (NGVD). The primary horizontal control is normally the State Plane Coordinate System. Sometimes, survey controls are already established on site from previous work or from known benchmarks, and they should be used for all project survey work, so that the existing and new features can be compared in the same space. However, in some cases, survey controls have to be brought in from off site, sometimes from many miles away, and this process can be time consuming and expensive. When the cost of establishing survey controls becomes expensive, as compared with the rest of the survey work and the size of the project, it would be justified to use an assumed local control for the work.

Local controls are based on an arbitrarily assigned vertical datum and coordinate system. When a permanent point is present at the site, such as a structure corner, property marker, section corner, or fence corner, it can be used as a local control monument, with an assigned elevation designation such as 100.00 or 1000.00, depending on the local relief, and an assigned coordinate datum, such as N10,000.00 and E10,000.00. All the survey work and the topographic map will then be based on this locally assigned control system. It should be noted that when the local control monuments are preserved and not disturbed, it is possible to eventually tie this local control system to the national and state system in the future.

3.6 Topographic Map

The finished product of a topographic survey is the topographic map. A topographic map is the starting element of any civil design project. The importance of an accurate topographic map in civil design cannot be overemphasized. There are many civil design projects

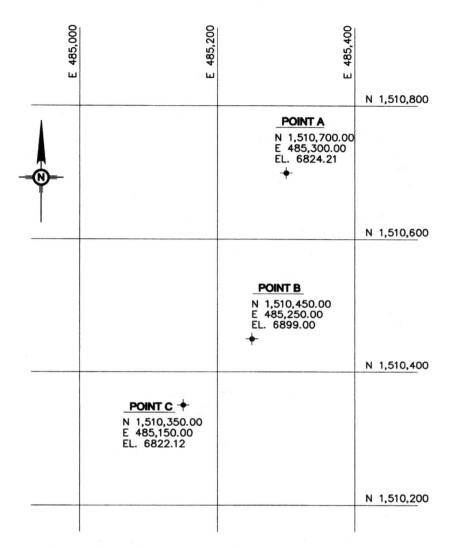

Figure 3-1. Horizontal coordinate system representation.

that required major redesign efforts because of problems and inaccuracies in the topographic base maps, particularly if the survey method is based on photogrammetry procedures. In one case, a major water conveyance project in East Africa was abandoned because of discrepancies and lack of coordination of separate topographic survey maps within the project area. It was discovered that the ground surface along the proposed downstream distribution canal was actually higher than the water-storage reservoir.

Regardless of whether the map is simple or complicated, there are basic elements that should be shown on a topographic map (see Figure 3-2):

- Topographic contours and contour labels. Typically, every five contour lines are labeled, and the labeled contour is called a major contour, and a heavier-weight line

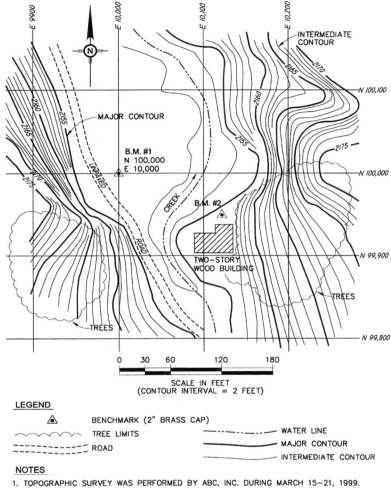

Figure 3-2. Typical topographic base map.

(see Section 7.7) is commonly used to distinguish it from the remaining contours that are called intermediate contours.

- Locations of control monuments within the mapped area, and the vertical datum and horizontal grid control system used.

- A north arrow, usually showing both the true north and magnetic north directions, with the current declination.

- A written scale and graphic bar scale.

- All surveyed and known natural and artificial features, such as trees, waterways, utilities, buildings, roads, culverts, and underground structure.

- The survey date and name of the surveyor.

- Legend of the features.

- Grid system of northing and easting.

- Spot elevations for specific high and low points.

Selection of the contour interval of a base map, which is the vertical distance between adjacent contours, should be the responsibility of the designer. In general, the contour interval is selected based on the purpose of the map, relative total relief of the project area, and required accuracy in the vertical direction. The contour interval should be specified as part of the survey contract requirements. A designer needs to balance the design requirements with the available project budget because the smaller the contour interval, the more expensive the survey. The American Society of Civil Engineers has guidelines for contour intervals for various projects (ASCE 1999), and these guidelines can be used as a starting point. At critical points, such as peaks, low spots, steep slopes, stream inverts, culverts, and highway crossings, enough spot elevations should be surveyed and shown on the map. These critical points should be specified by the designer and then picked up by the surveyor during fieldwork. It is important for the lead designer to perform a site visit to become familiar with site conditions and to reasonably evaluate the accuracy of the survey data submitted by the surveyor.

Topographic maps are now typically digitized electronically, and the electronic files of the maps can be used directly for civil design using one of many computer-aided drafting (CAD) software packages. Digitized maps can be plotted at any desired scale, and map scale becomes less of a surveying issue than it used to be. Regardless of what software is used, the contour lines of a digitized topographic map should be polylines, which are continuous three-dimensional lines of straight-line segments and contain the northings, eastings, and elevations (x, y, and z) information. Survey data in an electronic file of a topographic map are usually organized in layers. Some of the typical layers are surveyed points, contour lines, utilities, buildings, natural features (streams, trees, etc.), roads, and text and lettering. The establishment of individual layers can be specified as project requirements, but usually is left to the discretion of the surveyor. Each layer can be turned on or off electronically to yield a map containing different levels of information.

3.7 Levels of Investigation

Civil engineering projects are typically completed in phases or levels:

- Feasibility and conceptual levels—At these levels, various concepts, options, or alternatives are being proposed and evaluated, and the advantages, disadvantages, constructability, and approximate project costs are identified for each option. Also, there are no contract documents needed at these levels. It is also at these levels that fatal flaws are identified that would eliminate certain options from further considerations. If more than one site is being considered, then each site should be investigated on an equal basis for site selection evaluation. An adequate site investigation at this level should not be overemphasized. There is a misconception that an adequate site investigation is not needed at these early levels, but an adequate site investigation is not the same as a thorough investigation. At these levels, the site investigation should be sufficient to provide enough information so that the relative project cost for each option can be reasonably estimated without leaving any significant issues unresolved that may later "kill" the project because of cost.

- Final design level—At this level, a project site and a particular design concept have been selected for detailed design. It is at the final design level that contract documents are prepared. The site investigation at this level should yield the following information:

  1. All required design parameters, additional geologic investigation, subsurface investigation, and field and laboratory testing.

  2. Adequate topographic mapping so that an accurate design can be prepared and an accurate quantity estimate can be made. Typically, the coverage of the surveyed area and the contour interval are two important considerations at this level.

  3. Construction factors such as methods of excavation, excavation subgrade conditions, stability of temporary excavation slopes, dewatering requirements, etc., to establish a basis for bidding.

  4. If the design is performance-based (see Section 17.3 for a discussion of performance-based specification), then the investigation should contain adequate design and test data for the contractor.

PART 2

CONSTRUCTION DRAWINGS

CHAPTER 4

CIVIL DESIGN DRAWINGS

4.1   Introduction

Heavy civil engineering projects, such as highways, dams, and bridges, are constructed using documents consisting of drawings and specifications. The term *plans* is frequently used for construction drawings. Design drawings are graphical tools that designers use to depict their concepts and communicate the design concept to the builder (contractor). A typical set of plans used in heavy construction projects may contain elements from different engineering disciplines. These elements are divided into groups, such as *civil drawings, structural drawings, mechanical drawings, electrical drawings,* and *architectural drawings.* Although most of these groups can be distinguished readily from one another, the most ambiguous group is that of the civil drawings. For example, structural drawings are used for reinforced concrete and structural steel; mechanical drawings are used for mechanical equipment such as pumps, gates, pipes, and valves; architectural drawings are used for buildings, room layouts, elevators, etc. The definition of a civil drawing requires some discussion.

No readily available definition of a civil drawing exists. In the absence of any guideline, the following definition is proposed. A civil drawing shows the interrelation between structures and grades. The term *grade* is used to designate a surface. An *existing grade* denotes existing ground surface or contour before construction. An *excavation grade,* or *subgrade,* denotes a ground surface or contour created by excavation during construction. A *final grade* denotes the finished ground surface or contour after construction, and can be created by excavation and/or backfilling. By this definition, the following drawings would fall into the category of civil drawings:

- Plan of existing conditions, showing topography and surrounding features and structures;

- Plan of survey control, showing baselines, existing and new benchmarks, and coordinate grids;

- General plan of new structures, showing location and final grading around the new structures;

- Sections, profiles, and details of new structures showing existing ground surface, limits of excavation, locations of backfill, and final grades.

This chapter addresses the preparation of civil design drawings as defined by these criteria.

Preparation of civil design drawings requires knowledge and training in engineering graphics, descriptive geometry, and topographic survey. The recent development of high-speed personal computers allows this design process to be performed using various computer-aided drafting (CAD) software products. Chapter 11 is devoted to civil design processes using CAD systems. Knowledge of CAD skills is helpful, but is not necessarily an essential skill to provide a sound civil engineering design. An understanding of construction methods and constructability (the ability to actually erect the structure or work), gained through construction field experience and observing construction in progress, is important in the production of a sound civil design.

The level of detail in design drawings depends on the purpose of the drawings. The detail required depends on the levels of design (see Section 4.2). Drawings are used in conjunction with technical specifications for construction, and each contains different information (see Section 4.3), information that should not be repeated in both.

4.2   Levels of Design Drawings

Like most engineering projects, heavy civil engineering projects are developed in phases, or levels. A typical chronological sequence of phase development includes the planning phase, the feasibility or conceptual phase, the final design phase, and the construction phase. Each phase is associated with technical requirements and a corresponding level of project-supporting design documents; both are used to develop the project cost estimate. Therefore, a planning-level design is performed during the planning or permitting phase, a feasibility-level design is performed during the conceptual design phase, and a final design is performed during preparation of construction documents.

In general, the level of detail contained in the drawings increases as the project advances to subsequent phases. The following is a set of general guidelines for developing drawings at various phases:

Planning-level drawings. Planning-level drawings are developed and used at the beginning of a project to illustrate a particular concept or idea being considered. Typically, more than one concept or idea is being considered at this time, and sometimes these concepts lack supporting analysis or precedence, and therefore may not have sound technical bases. In fact, at this level, only limited engineering analysis is performed for each design concept, and most of the time, design drawings are developed based on project requirements and constraints, engineering judgment and experience, precedence, and past projects. Although often considered easier to prepare than final design-level drawings, the preparation of planning-level drawings—without the support of analysis and other design data—requires considerable design experience and judgment on the part of the designer. Therefore, inexperienced and junior design staff should perform this work only when closely supervised by more experienced designers.

Planning-level drawings should be developed with an equal level of detail for all concepts being considered. If no cost estimate is required to support each concept, then only sketches are needed to illustrate each concept, and key dimensions are not needed. If a cost estimate is required, then some key dimensions or elevations will be necessary to determine the quantities of materials needed. Planning-level concepts are typically evaluated and screened to a few favorable concepts for further evaluation in the conceptual-level phase. Other than those required to call out the types of general products or materials involved, no technical specifications are needed to support the drawings. In most cases, only a gen-

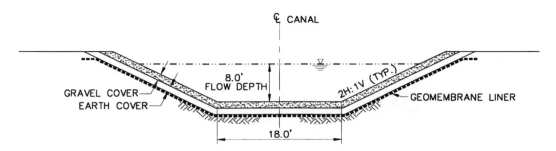

Figure 4-1. Geomembrane liner concept for a planning-level canal design project.

eral plan and a typical section are required to illustrate each concept. In all cases, planning-level drawings should be based on conservative assumptions.

An example of a planning-level drawing is shown in Figure 4-1. The figure shows a typical cross section of a geomembrane liner concept for a new canal project. The section shows the liner being protected by an earth cover and a gravel cover. The wetted surface (bottom width of the canal and the flow depth) has been determined based on project flow requirements; however, numerous design elements (the type of geomembrane material, thickness of the cover materials, and freeboard requirements) have not been determined at this time.

Conceptual-level drawings. Conceptual-level drawings for each project option are developed primarily to compare construction costs, to evaluate advantages and disadvantages, and to identify potential fatal flaws so that a preferred concept can be selected for final design. The evaluation criteria for various concepts may vary considerably from project to project, and may include cost, schedule, risk, and environmental, regulatory, and other nontechnical considerations. For many projects, construction cost estimates developed at this level are used as a basis for funding of the project, so it is important that sufficient cost is identified and based, in part, on the information furnished from the conceptual-level drawings. With this goal in mind, these drawings should be developed with sufficient details and dimensions to define all of the significant cost items associated with each concept. These drawings should not be used for construction, and not all the information needed for construction is shown on the conceptual-level drawings.

A plan and a typical cross section supplemented with relevant details will be required. In some cases, if a typical cross section does not represent the construction for the entire project, other cross sections will be required to obtain more accurate work quantities. Some engineering analysis may be required to support the design at this level, and the analysis performed at this level represents a refinement from the planning-level design. In general, with or without analysis to support the design at this level, the conceptual-level design should be conservative because of the following:

- The layouts, details, and sections developed are still inadequate to provide an accurate quantity estimate.

- Not all technical issues are addressed or anticipated until final design, when more detailed analyses are performed.

- The site may not be adequately investigated at this time to allow for consideration of all anticipated problems associated with subsurface conditions or other site constraints.

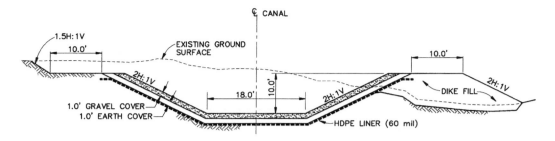

Figure 4-2. HDPE liner concept for a conceptual-level canal design project.

Figure 4-2 is an example of a conceptual-level drawing, shown as a refinement from the example in Figure 4-1, and illustrating the additional level of detail. The geomembrane-lined canal will be constructed using the cut-and-fill method. A 60-mil (0.03-mm)-thick high-density polyethylene (HDPE) liner was determined to be most suitable for this project. The proposed thicknesses of the earth cover and gravel cover are also shown on the section. The required freeboard will be 2 feet (0.6 m); therefore, the depth of the canal will be 10 feet (3 m). These dimensions will be used to determine the conceptual-design cost estimate for this option for comparison with other options; however, it should be noted that the details, geotechnical information, and survey data at this level are still inadequate for construction.

Final design drawings. Final design drawings, or *construction drawings,* are used for construction. These drawings contain all of the information necessary for a contractor to bid and build a particular project. Most civil engineering projects require permits from various regulatory agencies (municipal, county, state, and federal); these drawings are used to support applications for these permits. Construction drawings are used in conjunction with a set of technical specifications to define completely the spatial, material, and quality requirements of a project. Construction drawings are also used during final design to obtain an accurate estimate of quantities for an engineer's cost estimate and for use in developing the bidding schedule.

Figure 4-3 is an example of a construction drawing, shown as a refinement from the example in Figure 4-2, and illustrating the level of detail necessary for construction. Additional details include anchoring at the top of the HDPE liner, curved transition at the slope break at canal invert, stripping depth, permanent excavation slope, sloping on top of the service road and dike crest, and canal inverts at the section shown. Note that all material requirements, including the HDPE liner, are deliberately omitted from the drawings because these material requirements are now contained in the technical specifications and should not be repeated on the drawings.

Occasionally, bid amendments are used to change design drawings during bidding. After the contract is awarded and before construction begins, these design changes can be incorporated into the drawing set, and the revised set of drawings is referred to as *conformed drawings* (see Section 13.1).

When a construction project is completed, the construction drawings are updated and modified to reflect all of the changes made during construction. The resultant drawings are called *record drawings.* Record drawings (see Section 13.2) have been commonly referred to as *as-built drawings,* but the current trend is to use the term *record drawings.*

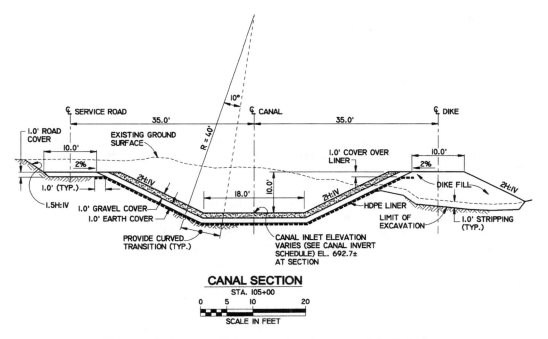

Figure 4-3. HDPE liner section for a canal final design.

4.3 Drawing Information

Unless otherwise stated, the remaining treatment of civil drawings will focus on final design drawings (construction drawings). Construction drawings, when combined with technical specifications, provide all of the information given to the contractor to construct and build a project. Information contained in construction drawings and information contained in technical specifications are intended to complement each other, and duplication between these documents should be avoided.

In general, civil construction drawings contain the following:

Location. Locations of existing conditions and new structures are shown on the drawings. Existing conditions include topography and surface features (e.g., vegetation, exposed structures, roads, streams). Buried structures and utilities, if known during design, should also be shown or noted on the drawings. Subsurface conditions, including soil types, bedrock, and groundwater level conditions, are also considered as existing conditions and should be shown. The locations of new structures or new facilities are shown in terms of plan, sections, profiles, and details.

Dimensions. Dimensions of new structures or new facilities, including lengths, angles, elevations, and thicknesses, are shown on the drawings.

Survey control and grades. Survey control and grades are shown on the drawings for construction survey of the new structures or new facilities. If existing benchmarks are located on site, they should be shown on the drawings. If off-site benchmarks are required for construction, their locations should be defined on the drawings. In fact, the survey basis of the project, whether it is based on local coordinates or state plane coordinates, local elevation or NGVD elevation, should be clearly stated on the drawings. All construction grades should then be based on the specified project control. Other project references, such as baselines, temporary survey points, permanent benchmarks, and existing structures not part of the work, should also be shown on the drawings.

Sequencing. Some construction projects require proper sequencing, and it may be possible to describe the sequencing in the specifications. If the engineer assumes a particular construction sequence or erection procedure, this assumption should be shown on a drawing. A graphical representation, supported by notes on the drawings, may be required to fully illustrate the sequencing requirements for a particular project.

Conflicts and interference. Conflicts and interference between new structures and existing facilities, or among various features of the new structure, frequently occur during construction. If anticipated during design, these problems should be shown on the drawings. Anticipated conflicts and interference not shown and not disclosed on the drawings or elsewhere in the construction documents will become grounds for change orders, claims, and disputes by the contractor.

Schedules and tables. Lists of design data that are used to define structure locations, curve data, sizes, dimensions, grades, etc., are frequently summarized in tables known as *schedules* on the drawings. These could include roadway centerline tangent and curve information; property line azimuths, lengths, and coordinates; and culvert sizes and invert elevations.

CHAPTER 5

BUILDING A SET OF CONSTRUCTION DRAWINGS

5.1 Drawing Sheet Size

Drawing sheet size refers to outside-to-outside dimensions. In the United States, there are two standard drawing sheet sizes in the engineering design industry for full-size drawings: 24 in. × 36 in. (610 mm × 914 mm), and 22 in. × 34 in. (559 mm × 864 mm). The predominant size is the former. The architectural design profession also uses a larger 36 in. × 48 in. (914 mm × 1219 mm) sheet, but this size is rarely used in engineering design. Based on historical precedence, some owners and state and federal agencies use drawing sizes that are somewhat different than the standard sizes. For example, the standard sheet size for the U.S. Fish & Wildlife Service Division of Engineering construction drawings is 21.5 in. × 36 in. (546 mm × 914 mm), and the basis for this size is unknown.

The size of the drawings is selected at the beginning of the design. The selection may be based on client requirements or, in the absence of client requirements, the designer's general preference. Recently, designers and owners have used half-size drawings during reviews and submittals. Half-size drawings have the following advantages over full-size drawings:

- They are less bulky and can easily fit into briefcases and file cabinets.

- Because they do not require the Diazo blueprint process or a large-format copy machine, they can be copied easily and inexpensively in a conventional copy machine.

- They are preferred by contractors for ease of use in the field.

In a true half-size drawing, all the scales are reduced to exactly 50%, and therefore, dimensions can still be scaled where needed. For a 24 in. × 36 in. (610 mm × 914 mm) drawing, the half-size drawing sheet is 12 in. × 18 in. (305 mm × 457 mm). For a 22 in. × 34 in. (559 mm × 864 mm) drawing, the half-size drawing sheet is 11 in. × 17 in. (279 mm × 432 mm). Nearly all conventional copy machines can provide 11 in. × 17 in. copies, but making 12 in. × 18 in. (305 mm × 457 mm) copies requires a special, large-format copy machine that is more expensive. For this reason, some designers select a 22 in. × 34 in. (305 mm × 457 mm) full-size drawing so that true half-size drawings can be easily copied. For designers that prefer a 24 in. × 36 in. (610 mm × 914 mm) sheet that offers a larger drawing space, a near half-size drawing can be obtained by copying on 11 in. × 17 in. (279 mm × 432 mm) sheets, except the scale is reduced to about 47% instead of 50%. It should be noted that a 24 in. × 36 in. (610 mm × 914 mm) drawing prepared in computer-aided drafting (CAD) (see

Chapter 11) can be plotted in true half scale directly on an 11 in. × 17 in. (279 mm × 432 mm) sheet, provided that there is sufficient space in the borders for margins.

5.2   Drawing Title Block

After the drawing sheet size is selected, the drawing title block information is assembled. The title block format of a construction drawing is usually fixed for a particular design firm or government agency, and should not be changed from project to project unless specifically requested by the client. Many different title blocks have been used, and standardization of the title block is not intended here. Most title blocks are located on the bottom or right side of the sheet. Regardless of title block's location on the sheet, it should contain the following:

Project identification. Project identification should include a brief title of the project and the project location (city, county, and state).

Sheet identification. Each sheet should be properly numbered and identified with the full title as listed on the cover sheet.

Revision number/block. Revision numbering is the designer's way to keep track of changes made to a particular drawing or the entire set. Some designers prefer to use a numbering system that starts at Revision 0, and progresses to Revision 1, 2, etc., until the design is complete and ready for bidding and construction. Others prefer to use lettering (A, B, C, etc.). Either method is acceptable. The revision block is used to describe the reason for the revision, and the block contains the revision number, description of the revision, date of revision, and usually the principal designer's initials. There are many reasons to change a design during production: progress submittals for reviews, changes in design criteria and project requirements, etc. Some designers prefer to switch revision designations (e.g., from lettering to numbering) when the design is complete and the project enters into the bidding and construction phase, with Revision 0 designated as "issue for bid" or "issue for construction," and Revision 1, 2, 3, etc., for later revisions during construction.

Designer and owner identifications. The logos, names, and addresses of the design firm and owner of the project should be shown on the title block.

Responsible personnel identification. The initials or signatures of responsible personnel are part of the title block. Different designing entities have different project personnel classifications and designations, and they also have different policies as to whose names or initials can be put on the drawings. In general, the following lead design personnel categories are contained in the title block:

Designer: The *designer* is responsible for the technical content of a particular drawing. Depending on the complexity of the project, there may be more than one designer for a given set of drawings. For example, one person might be in charge of geotechnical design while another person is in charge of structural design. The designer may or may not be a licensed professional engineer, depending on the subject matter of the design.

Drafter: The *drafter* is responsible for drafting production of a particular drawing. Essentially all modern engineering design is prepared in CAD; when a designer prepares the design in CAD, the drafter may also be the designer.

Checker: A *checker* is responsible for quality control during production of a drawing. His or her responsibilities includes back-checking drawings after they are drafted to assure compliance with drafting standards and the design's technical intent. In small design firms, the checker and the designer may be the same person. In cases in which the drafter is the designer, it is important that the checker and the drafter are not the same person. In large firms, a separate checker is usually assigned to a project for the sole purpose of quality control.

Reviewer: A *reviewer's* responsibility is to review the design for compliance with all technical requirements of the project. The reviewer should be an experienced practitioner who has design experience and is technically qualified for the project. Whereas the checker is involved during production of the drawings, the reviewer is involved only at milestone completion stages before the design documents (plans and specifications) are submitted to the client. It is important that the reviewer focuses primarily on critical issues (e.g., constructability, technical solutions, and value-engineering issues to decrease costs), and not just drafting or editorial comments, which could cause him or her to miss the essential intent of the reviewing process.

Project manager: In the engineering consulting profession, the *project manager* is the person in charge of a particular project, whether administratively or technically, or both. Depending on his or her background and the size of the project, a project manager may also be the principal designer or the reviewer.

Engineer's seal. Construction drawings are stamped and signed by a licensed professional engineer. There are no guidelines on the location of the engineer's seal. In most cases, the seal is located at the bottom of the sheet. Some title blocks contain a square space that is reserved for the engineer's seal.

Sometimes, drawing scales are shown on title blocks. This practice may lead to confusion and is not recommended, as a drawing may contain more than one scale.

Figure 5-1 shows an example of a title block at the bottom of the sheet. Alternatively, this layout can be rotated vertically and placed on the right side of the sheet.

5.3 Sheet Organization

There should be a logical order to the organization of the construction drawing sheets, regardless of the size of the design package. When a logical order is followed, the organized manner allows the design team to divide up and produce the drawings according to various design disciplines, and the contractor benefits from the ease of obtaining needed construction information and passing the information down to his or her subcontractors and supervisors.

The following sequencing of drawings is commonly used:

1. Title Sheet

2. Abbreviations/Legend/General Notes

3. Existing Conditions

4. Subsurface Conditions

5. Group Drawings (Civil, Structural, Mechanical, etc.) for each group
   General Layout Plans
   Detailed Layout Plans
   Sections and Details Sheets

6. Miscellaneous Details Sheets

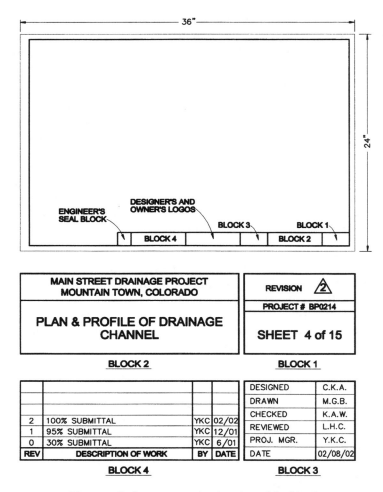

Figure 5-1.   Example drawing title block.

The principle of sheet organization is simple: the series of sheets should be organized from general to specific. After the title sheet and other general sheets that define the existing conditions, the remaining sheets should show the general structure layout and sections, followed by more specific layouts and details. In heavy civil construction, the general civil drawings will precede drawings from other disciplines, which are organized into groups. Where the work does not seem to fit into any well-defined categories, these design features are grouped into the so-called miscellaneous details at the end of the package.

The following is a discussion of the general requirements and guidelines for preparing various parts of a set of construction drawings:

Title sheet. The title sheet, of course, is always the first sheet for the drawing package. It is also referred to as the *cover sheet.* The cover sheet contains the following information:

- Name of the project, usually in very large and bold letters at the top portion of the sheet.

- List of drawings contained in the package. For a large package, the list of drawings cannot fit on the cover sheet, and is usually located in the sheet immediately following the cover sheet.

- Location maps of the project in relation to highways, access roads, and nearby towns and cities.

Abbreviations, legend, general notes. These sheets are used to define the abbreviations used in the drawings, graphic symbols/legends, and general notes that apply to the entire drawing set.

Existing conditions. Several drawings may be required to define the existing (preconstruction) conditions of the project site. Existing conditions to be defined include a topographic map of the site, survey controls, subsurface conditions, utilities, borrow areas, and other information obtained from site characterization (see Chapter 3). Some designers prefer to present climatic and meteorological data of the site in a graphical manner on the drawings (see Section 20.2).

Subsurface conditions are generally presented on the drawings in terms of a plan of exploratory boreholes and test pits, and simplified graphic logs of boreholes and test pits can be used to summarize the results of these investigations. More specific results of subsurface conditions can be found in geotechnical reports or as reference data incorporated into the technical specifications (see Chapter 20).

Some projects are so large that multiple sheets are required to show the topography. In such cases, key maps are required to show the relative locations of sheets, and match lines (see Section 7.7) are used to show the points of overlap of the drawings.

General plan and sections. The general plan or overall plan of the new feature should be a civil drawing showing the finished grades or contours. Section cuts from the general plan should show existing ground surface, limits of excavation, and fill limits. All required grades for the project should thus be shown on these sheets.

Detailed plan and sections. A more-detailed layout plan and sections of structures may be required. If a general plan and sections have already been used to show all required grades, then the detailed plan and sections can show only details pertaining to the structure itself, without surrounding topographic information to simplify these drawings.

Grouping of disciplines. Heavy civil construction projects typically include multidisciplinary work such as structural, mechanical, and electrical components. Design packages of these features are generally grouped into sheets of the same discipline. Therefore, after the civil drawings, one or more groups of drawings can follow. To represent drawings of different disciplines, some design firms use prefixes on the sheet numbers. The following is a common set of prefixes for disciplines encountered in heavy civil work:

G: General drawings (cover sheet, list of drawings, abbreviations, legend, general notes, existing site conditions, etc.)

C: Civil drawings (general plans and sections, excavation plan and sections, earthwork, etc.)

S: Structure drawings (reinforced concrete, structural steel, metalwork, etc.)

M: Mechanical drawings (gates and valves, piping plan and details, pumps, etc.)

Sheet Number	Description
G1	COVER SHEET, LOCATION MAPS, LIST OF DRAWINGS
G2	GENERAL NOTES, ABBREVIATIONS, & LEGEND
G3	GENERAL SITE PLAN, BORROW AREAS, & STAGING AREA
G4	EXISTING PLAN OF DAM & SPILLWAY
G5	BOREHOLE LOGS & TEST PIT LOGS
G6	GENERAL PLAN OF DAM MODIFICATIONS
G7	RECLAMATION – PLAN & DETAILS
C1	PLAN OF EMBANKMENT MODIFICATIONS
C2	PLAN OF EXCAVATIONS
C3	EXCAVATION SECTIONS & DETAILS
C4	EMBANKMENT – SECTIONS & DETAILS, SHEET 1 OF 2
C5	EMBANKMENT – SECTIONS & DETAILS, SHEET 2 OF 2
C6	EMBANKMENT – INSTRUMENTATION SCHEDULE & DETAILS
S1	PLAN OF SPILLWAY MODIFICATIONS
S2	SPILLWAY PLAN & SECTIONS
S3	SPILLWAY – SECTIONS & DETAILS, SHEET 1 OF 2
S4	SPILLWAY – SECTIONS & DETAILS, SHEET 2 OF 2
S5	SPILLWAY REINFORCEMENT DETAILS
S6	PLAN OF OUTLET WORKS MODIFICATIONS
S7	OUTLET WORKS – SECTIONS & DETAILS
S8	OUTLET WORKS REINFORCEMENT DETAILS
M1	OUTLET WORKS GATE & VALVE DETAILS

Figure 5-2.   Example of list of drawings.

E: Electrical drawings (electrical site plan, wire diagram, conduit diagram, one-line diagram, details, etc.)

A: Architectural features of buildings (floor plan, windows and doors, finishes, etc.)

Miscellaneous detail sheets. Miscellaneous detail sheets are used for details that do not fit logically into any of the special disciplines listed above. Typical details shown on miscellaneous detail sheets include standard fencing, instrumentation, small fabricated metalwork, survey monuments, and temporary or permanent signs. These sheets typically represent the last sheets in a drawing set.

Figure 5-2 is an example of a list of drawings for a dam project, illustrating the sequencing of drawings from different groups and disciplines. It is important to point out that no two drawings should have the exact same title. In the example shown in Figure 5-2, there are two sheets for *Spillway—Sections & Details* and two sheets for *Embankment—Sections & Details*; these two sheets are differentiated by adding "Sheet 1 of 2" and "Sheet 2 of 2."

CHAPTER 6

LAYOUT OF A CIVIL DESIGN PLAN

6.1 Design Controls

The layout of a civil design requires a set of design control references that are used (1) to determine the design alignment and dimensions, (2) to avoid interference between different members of the design team, and (3) to enable the construction surveyor to establish the line and grade to build the project.

Design controls for civil design work consist of benchmarks, coordinate grids, control points, project baselines, and centerlines of structures. To account for ease of layout during design and ease of survey during construction, the selection of locations of design controls should be considered carefully.

Control points. Control points may be primary survey benchmarks (see Section 3.6) established on site prior to construction, and/or additional control points used to establish project baselines and centerlines. Primary survey benchmarks should contain information on elevation and coordinates (northings and eastings). Control points that are used to define a baseline or centerline only require information on coordinates and stationing (see Section 6.2). They are established by the contractor during construction, and they may only be temporary working points. All control points should be located beyond the limits of excavation and backfill operations so that they will not be disturbed during the work. They should also be recorded on record drawings to make project records more understandable when referenced.

Baseline/centerline. Baseline is a linear or nonlinear reference to establish project stationing and from which offset distances are taken. Centerline is usually associated with a particular structure or design feature, such as an embankment, canal, pipe, or highway. A centerline can also be a baseline. A minimum of two control points, or one control point plus a bearing (true north, magnetic north, or azimuth), are required to establish a straight baseline or centerline. A baseline or centerline can also be a curve. Any changes in bearing for a baseline or centerline should be noted with either the new bearing, or the deflection angle.

Figure 6-1 is an example of a set of design controls for a dam raise project. Benchmark numbers 1 and 2 are used for primary survey controls. The benchmark designations, coordinates, elevations, and stations are shown on the figure. A baseline is established by these two benchmarks. This baseline is also the centerline of the existing dam crest. The dam is

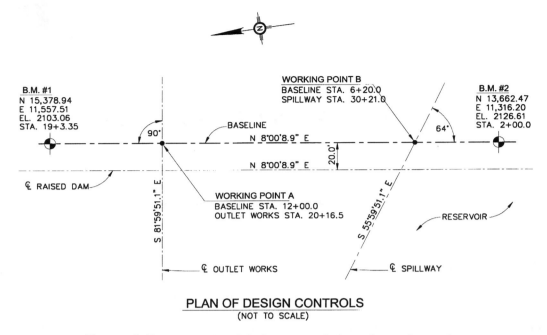

PLAN OF DESIGN CONTROLS
(NOT TO SCALE)

Figure 6-1. Example of design controls for a dam raise project.

raised by placing fill upstream, and the centerline of the raised dam crest is parallel to the baseline and offset a distance of 20 feet (6.1 m) upstream of the original baseline. A new outlet works is constructed by a cut-and-cover method in the middle of the dam. The centerline of the outlet works is established by a bearing and Working Point "A." A new spillway is constructed in the right abutment of the dam. The centerline of the spillway is established by a bearing and Working Point "B."

6.2   Stationing and Offsets

In a civil design plan, any given point on the plan can be represented by the project coordinates (northings and eastings), which are cumbersome, and should be reserved only for key control points. When the design controls include baselines and centerlines, any point on the plan can be referenced to these lines by means of stations and offsets. Stations are distance labels along a baseline or centerline, similar to street addresses. The standard designation for stationing is $x + z$, where x is an integer including zero, and z is a two-digit number between 00 and 99. The integer x is used to express distance in numbers of hundreds of feet (or hundred of meters in the metric system). When the value of z is less than 10, it is written with two digits (e.g., 02). For example, station $5 + 07.1$ is a point on the centerline that is 507.1 feet (154.56 m) from an arbitrary zero reference point (station $0 + 00$). When a point on the centerline falls behind the zero reference point, that point has a negative stationing. Negative stationing is used by some designers, but this practice should be discouraged to avoid confusion associated with the negative sign. To avoid using negative stationing, station $0 + 00$ should be well beyond the project limits. For example, the starting stationing for a project can be taken as $10 + 00$ or $100 + 00$, depending on the size of the project. Stationing is typically labeled at some regular interval along a baseline or centerline, such as every 1, 5, or 10 stations (100, 500, or 1000 feet), depending on the length of the line.

When a project contains more than one baseline or centerline, different stationing may be required. When different stationings are used for a project, each stationing should be identified and differentiated. In the example on Figure 6-1, three stationings are used:

1. Stationing for the baseline and centerline of the raised dam. The same stationing can be used for these two lines because they are parallel to each other. The stationing is called out on the drawings as *BASELINE STA*.

2. Stationing for the outlet works structure. The stationing is called out on the drawings as *OUTLET WORKS STA*.

3. Stationing for the spillway. The stationing is called out on the drawings as *SPILLWAY STA*.

In addition, when more than one stationing is used for a project, it is advisable to use different numbering systems to make it easier to differentiate them. In Figure 6-1, the dam baseline station is from 2 + 00 to 19 + 03.35. The outlet works station can be from 20 + 00 to 30 + 00, and the spillway station can be from 30 + 00 to 40 + 00. In this way, a station designation of 25 + 00 will imply it is on the outlet works centerline, and a station designation of 15 + 02 will imply it is on the dam baseline.

Offsets are distances measured perpendicular to the baseline or centerline. On the plan, the offset for a particular point can be dimensioned directly from the point to the line. Frequently, a table is used on the drawings for a series of offset points on both sides of the centerline. To denote which side of the centerline a point falls on, a sign convention is sometimes used, with positive offset on one side, and negative offset on the other. Positive offsets should be to the right, looking upstation. Alternatively, subscripts r and l are added to the offset, again looking upstation.

6.3 Scale Selection

The selection of a proper scale for a civil design plan is controlled by the amount of paper space reserved for that plan. In general, the proper scale is selected to allow the maximum amount of information and details to be shown on the drawing space available. Because essentially all of the civil design drawings are now prepared in some form of computer-aided drafting (CAD) systems (see Chapter 11), this guideline requires some clarification. With CAD, any scale can be used for the drawing, because the software allows the base map to be enlarged and zoomed to the necessary scale. The proper scale in a CAD drawing refers to the scale that will be plotted on a full-size drawing sheet, which will be read and interpreted for construction by the contractor.

In civil design, the so-called civil engineer's scale is used. In the United States, where U.S. customary units are still commonly used in engineering design, the following scales, available from a standard civil engineer's scale, should be used: 10, 20, 30, 40, 50, and 60. Other scales (15, 25, 70, etc.) are not recommended because scaled distances cannot be readily obtained directly from the scale. When a 10-scale is used, it would mean 1 inch = 10 feet; a 100-scale means 1 inch = 100 feet.

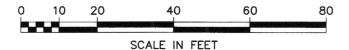

SCALE IN FEET

SCALE IN FEET

Figure 6-2.   Example of bar scales.

For detailed structural or mechanical design, the architect's scale is used. In the United States, the following scales are available from a standard architect's scale in U.S. customary units:

3-scale (3 inch = 1 foot)

1.5-scale (1.5 inch = 1 foot)

1-scale (1 inch = 1 foot)

¾-scale (¾-inch = 1 foot)

½-scale (½-inch = 1 foot)

⅜-scale (⅜-inch = 1 foot)

¼-scale (¼-inch = 1 foot)

³⁄₁₆-scale (³⁄₁₆-inch = 1 foot)

³⁄₃₂-scale (³⁄₃₂-inch = 1 foot)

In all cases, the scale should be selected considering the complexity of the plan or detail, and the legibility when half-size drawings are printed.

6.4   Scale Display

Bar scales with scale units should be used for all drawings. With the increased use of reduced drawings, a narrative description of the scale on the drawing (e.g. 1″ = 200′, ½″ = 1′–0″) does not adequately convey that information, because the drawing could have been photographically or electronically reduced to an unknown percentage of its original size. With a bar scale, the scale itself will be proportionally reduced as the drawing is reduced, thus eliminating the ambiguity. Figure 6-2 illustrates the use of bar scales.

CHAPTER 7

GRAPHICAL REPRESENTATION OF CIVIL DESIGN

7.1 General

This chapter outlines the graphical views by which civil drawings are represented. Typical views include *plan views* (or *plans*), *sections*, *elevations*, *profiles*, and *details*. These two-dimensional views are used to represent a three-dimensional picture, such as that of an open excavation, earthfill structure, or site grading. The contractor uses these views to establish the survey controls for lines and grades required to build the structures. These graphical representations are also used to estimate payment quantities for completed work (see Chapter 22). Unlike machine parts, three-dimensional views such as isometric or oblique views are too complicated for heavy civil construction, and are not practical to show on construction drawings.

Graphical views are assembled using lines and call-outs. The necessary line types and line weights will be described and illustrated in this chapter. For civil design, definitions and meanings of various line types are based on traditional practices of engineering and architectural graphics, but they are explained specifically for civil design applications.

7.2 Plan View

The plan layout of a civil drawing is the most important view of all graphical representations; all other views are supplemental or auxiliary to the plan view. The plan view of a civil layout drawing, regardless of structure type, should contain the following:

- A title.

- A bar scale.

- A north arrow, showing the orientation of the layout with respect to the compass direction (true north or site grid north).

- Base topographic contours, showing the existing contours both outside and inside the new structures. The existing contours under the new structures should be shown in dashed lines (see Section 9.7), because the new structures would alter the existing topography.

- Project coordinate grid lines, expressed in terms of northings and eastings. These grid lines should be labeled at regular intervals, such as every 100 feet (30.5 m) or every 250 feet (76.2 m), depending on the size of the project. Grid lines are sometimes shown on 2-inch (51-mm)-long lines on the borders, or cross-tics in the interior of the plan.

- Stationing of centerlines of major features, such as a roadway, dam crest, or a pipeline. The stations should be labeled at regular intervals, such as every 50 feet (15.2 m) or every 100 feet (30.5 m), depending on the size of the project.

- Section cut lines, showing the locations where cross sections are being taken. The section cuts should include a "bubble" symbol (see Section 8.1) showing the section designation, the sheet number where the section is cut, the sheet number where the section is shown, and the sequential designation of the cross section.

An example of a typical plan view is shown in Figure 7-1.

All exposed key elevations and slopes of the new structure should be called out on the plan. The exposed and buried limits of the structure should also be shown. The intersections of cuts or fills with existing ground are known as *catch points*. The trace of catch points is called a *catch line*, and both catch points and catch lines are established graphically. Most computer-aided drafting (CAD) programs (see Chapter 11) have capabilities to establish these catch points and lines without manual operations. However, it is important that

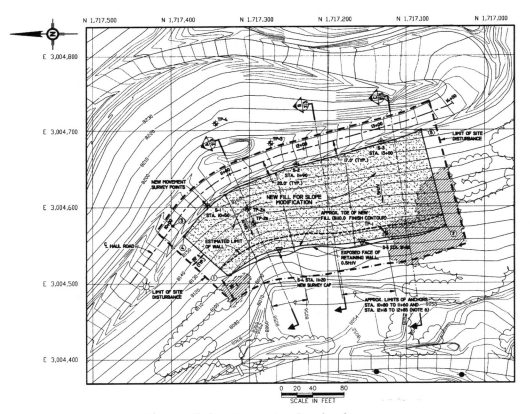

Figure 7-1. Example of a plan layout.

one should know the principles of manually establishing catch points, even though a computer would be used during production of the plan. The principles of establishing catch points and catch lines are explained in Section 9.2.

The plan view should also show the centerlines and/or outlines of major buried features, such as pipes, manholes, or drain blankets.

There are two methods to represent lines and grades. One method is to contour the entire feature; the second method is to use slopes and elevations without contours. Both methods will accomplish the same goal, namely, telling the contractor how to build the structure. Figures 7-2 and 7-3 illustrate the two methods of representing lines and grades. Note that the method used in Figure 7-2 requires significantly more lines to define the pond with new contours. Catch points are shown as black dots in Figure 7-2. Note that the same catch lines are used to define the outline of the pond in this example. Some designers prefer to use the second method, that is, using slopes and elevations, because it gives a cleaner drawing. Note that in Figure 7-3, the slope lines are perpendicular to the contours shown in Figure 7-2.

In general, the orientation of a feature on a plan is selected based on the best way to fit it on a drawing sheet, rather than on compass orientation. For example, although it should be the first choice, it is not always necessary to orient a feature so that the north direction points to the top of the drawing sheet (see Figure 3-2). It is a good practice, though, to be consistent in orientation for various portions of the plan view of the same feature in the drawing set. Many state highway departments have guidelines on plan orientation for roadways. Generally, the plan of a roadway is oriented so that the station increases from left to right on the drawing sheet.

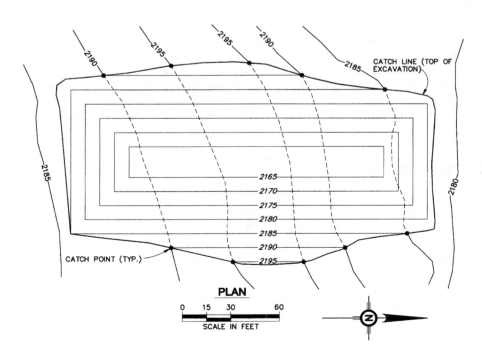

Figure 7-2. Plan of pond excavation shown using new contour method.

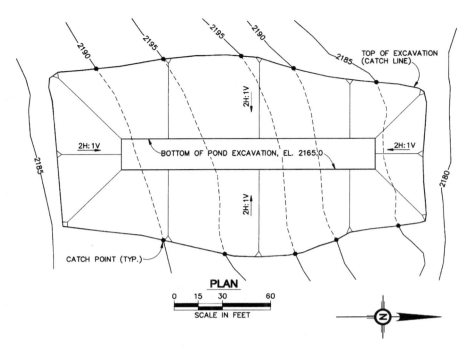

Figure 7-3. Plan of pond excavation shown using slopes and elevation method.

There is one particular type of feature for which the feature orientation on a plan view is controlled by stream flow direction. Dams are usually located on a stream (although there are off-stream dams), and the plan view of a dam should be oriented on the drawing sheet so that the stream flows from the bottom to the top of the sheet. The reason for this preference is that abutments of a dam, by convention, are referred in the dam engineering profession as the left and right abutments when one looks downstream. When a dam plan is oriented this way, the left abutment is on the left side of the drawing, and the right abutment is on the right side. Otherwise, one has to mentally or physically rotate the drawing in order to orient with reference to the abutments and viewer's location.

7.3   Section View

A section view is also called a *cross section*. A section shows the internal relationship among various components and features of a structure shown in the plan view. An adequate number of sections should be drawn to show all of the internal features and to construct the structure. For a linear structure, such as a buried pipe, sometimes a single typical section (Figure 7-4), may be sufficient to construct the entire pipe. For other more complicated structures, more than one section may be required to show all of the design information and relationships of the project elements.

A section should contain, as a minimum, the following:

- An elevation scale, preferably on both ends of the section, with the datum and units indicated on the scale.

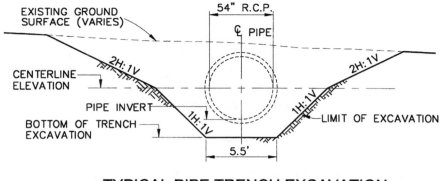

TYPICAL PIPE TRENCH EXCAVATION
(NOT TO SCALE)

Figure 7-4. Example of a typical pipe section.

- A title and section bubble (see Section 8.1) that is referenced to the drawing where the section is cut.

- A scale bar under the title. If no scale exaggeration is used, the same scale as the elevation scale is used here. However, if the section is exaggerated (see Section 9.5), then both the vertical and horizontal bar scales should be indicated here.

- Existing ground surface as obtained from the base topographic map. This information is important because it is a reference for cut and fill quantities, and it provides the contractor with an orientation of the new structure relative to the existing features.

- Limit of excavation, which may also be the bottom of the structure. The excavation is between the existing ground surface and the limit of excavation. The limit of excavation (also known as *excavation neat line*) is important design information that is also used by the contractor and owner for measurement and payment purposes. Excavation beyond this neat line is considered overexcavation, which requires special contract provisions for payment.

- Final grade (also called finish grade).

To minimize the number of sections required, the designer should determine the section location. The number of cross sections should be selected based on the following:

- Each section should show the maximum design information. When changes occur from one location to another on the plan, additional sections will be required to show these changes.

- Sections are required for quantity takeoff. Frequently, one section does not show a typical condition across the entire structure, and additional sections are required to define more accurately the quantities involved (see Section 22.3).

Sections can be labeled alphabetically (Section A, B, C, etc.). Because of potential confusion of the letters *I* and *O* with numbers, Sections I and O are generally not used. For a

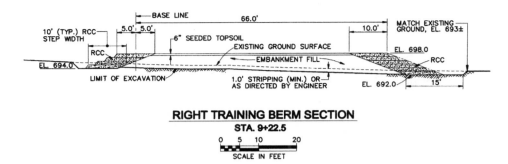

Figure 7-5. Example of a cross section at a specific location.

large design package, the 24 section designations are not enough for all of the required sections. In those cases, there can be more than one of each section in the drawing package.

Some designers prefer to label sections numerically instead of using letters. This approach allows unlimited sections to be used without repeating. In a large design package that includes general, civil, structural, and mechanical features, the sections should be numbered consecutively in each discipline. For example, sections in a civil package are numbered C1, C2, C3, etc., and sections in a structural package are numbered S1, S2, S3, etc. (see Section 5.3). When sections are designated by numbers, details should be designated by letters to avoid confusion (see Section 7.6).

When a section is cut at a special location, such as a certain station, that station should be indicated under the section title. Figure 7-5 is an example of a section view.

When a section shows several different types of materials, such as concrete, sand, gravel, and other types of earthfill, material symbols should be used to differentiate these materials and enhance the graphic quality of the drawing. Figure 7-6 illustrates the use of graphic symbols to enhance the drawing.

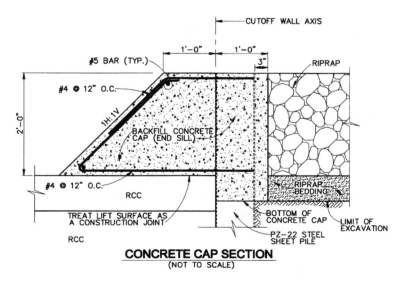

Figure 7-6. Use of graphic symbols to enhance a drawing.

7.4 Elevation View

Elevation view, as the name implies, shows the vertical relationship of various features of a structure. It is an external view of the side of a structure. This view should not be confused with a cross section, which cuts through a structure. If a structure is nonsymmetrical, various elevation views (north elevation, east elevation, etc.) may be required to show all of the features. An elevation view can also be a profile, but an elevation view should not have exaggerated scales.

Elevation views are frequently used for concrete and mechanical structures, but they are not typically used for earthwork structures. Earthwork structures are typically simple enough that plans and cross sections are adequate to show all of the design requirements. When a structure is viewed from the outside, all of the external components that are exposed are projected onto the same plane. In a sense, the elevation view is similar to the plan view, except the plan is a view from above, whereas the elevation view is a view from the side.

An elevation view should contain, as a minimum, the following:

- A title and elevation bubble that is referenced to plan or section views for which the view is taken.

- A scale bar under the title. No scale exaggeration should be used for the elevation view.

There is much flexibility and designer's discretion that can be used in showing the elevation view. Options that have been used include:

- An elevation scale, with the datum and unit indicated on the scale. Whether an elevation scale is used or not, key elevations of the structure should be called out (e.g., bottom of excavation, material boundaries, etc.).

- The compass orientation of each end of the structure may be helpful when various elevation views are taken for a nonsymmetrical structure.

- When an elevation view shows an isolated structure for simplicity, it is not necessary to show the adjacent features around that structure. However, if the view is intended to show, for example, the structure foundation, then the existing ground surface and the finished ground surface should be shown on the view as well. The information on the foundation earthwork should be kept to a minimum and as simple as possible because the key foundation earthwork design should be shown on a separate cross section.

An example of an elevation view is shown in Figure 7-7.

7.5 Profile View

A profile view is essentially a long cross section along an entire structure element. The main reason for using a profile is to show the entire structure longitudinally in a view that includes all of the associated structures and interrelated controls. Examples of structures for which the profile view is very useful in design drawings include dams, highways, sheetpile walls, slurry trench walls, tunnels, spillways, outlet works, and pipelines.

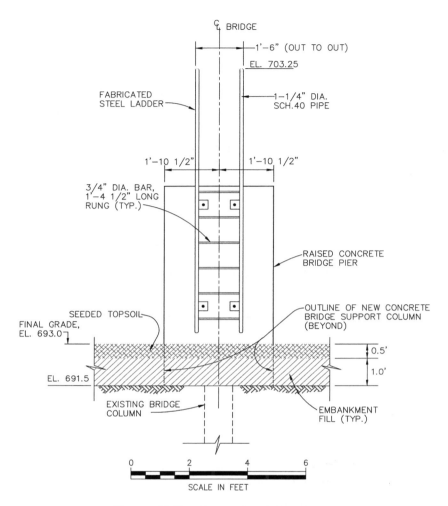

Figure 7-7. Example of elevation view.

Because the profile view is a cross section, all of the drawing requirements listed in Section 7.3 still apply. In most cases for long structures, it is necessary to exaggerate the vertical scale (see Section 9.5) to obtain good resolution of the design features. Figure 7-8 shows an example of a profile view.

7.6   Details

Details show the design features and requirements when the overall plan, sections, and other views do not have the proper scale and resolution to do so. Details are essentially enlargements of certain design features, and can be shown in any view that is convenient. Examples for which details are required include toe drains, weep drains, structural connections, pipe connections, and instrumentation. Typically, details are called out from other views, and design requirements are shown on these enlarged details. When showing the details, it is important that the same perspective be used. For example, if a certain portion of

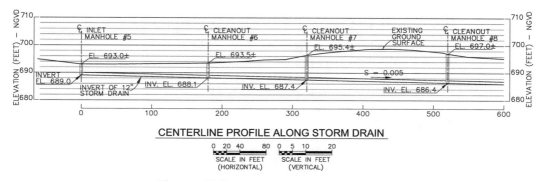

Figure 7-8. Example of profile view.

a plan view requires a detail, then the detail should also be in plan view, not in a section view. In addition, the detail itself should be oriented in the same way, and not rotated to other directions.

In addition to detail titles, details can be numbered consecutively (1, 2, 3, etc.) for each design discipline. Alternatively, letters should be used to designate details if a numbering system is used for cross sections (see Section 7.3).

When the purpose of a detail is to add dimensions and other callouts, it is frequently not necessary to draw the detail to scale. When a detail is not drawn to scale, it should be labeled as "NOT TO SCALE" or "NO SCALE" under the title (see Section 9.5). Examples of details are shown in Figure 7-9.

7.7 Line Types

A civil design drawing is composed of line segments that describe and define the geometry of specific features. Contrary to an artistic drawing, line segments in a technical drawing, such as a civil design drawing, have specific meanings and functions. Many line types are required to show a three-dimensional representation on a two-dimensional drawing sheet. Correct usage of various line types is essential in showing the features to be constructed. Incorrect applications of various line types can, in some cases, lead to misleading information and representation, and ultimately can cause confusion that requires clarification during construction. In this chapter, various line types that are commonly used in civil drawings are defined and described, and examples are provided to illustrate their applications. A summary of the line types and uses is presented in Figure 7-10.

Visible Line

A visible line (also called an *object line*) is a solid line that defines the outline, size, and shape of a feature that is directly visible or exposed in a particular view. It is the most basic building component of a technical drawing, and it should be the most prevalent and prominent line type in a civil design drawing. This line can be a straight line or any irregular shape that is required to describe a particular feature. A view is most effective and well-chosen when visible lines are used most frequently for that view. If a view contains too many other line types (e.g., hidden lines or phantom lines), then perhaps other views are more logical to show a particular feature.

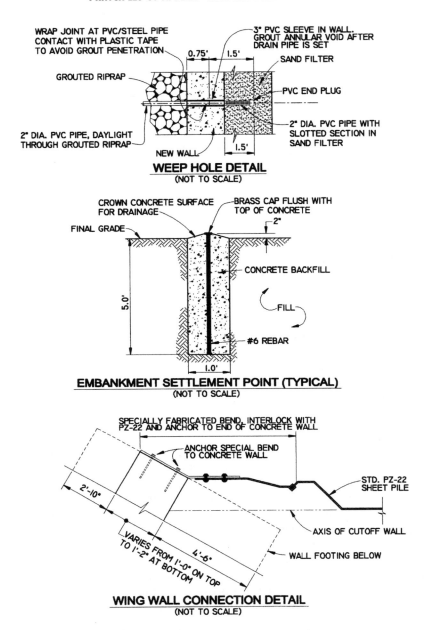

Figure 7-9. Examples of details.

Centerline

A centerline generally shows the line of symmetry or relative location of a particular feature. Examples will be the centerline of a roadway, dam crest, pipe, tunnel, or a drainage channel, as shown in Figure 7-11. A centerline is a very important reference, not only to prepare construction drawings of a particular feature, but it is also important to a contractor during construction of that feature. Typically, a layout of that feature starts with the centerline, and

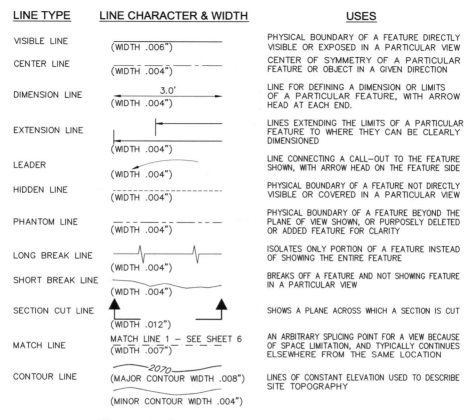

Figure 7-10. Summary of line types and uses.

dimensions are then taken from that line to draw the outline. Where two features intersect, the relationship between the two features (e.g., the station and angle of intersection) is best defined using the intersection of the two respective centerlines.

Dimension Line and Extension Line

A dimension line defines the distance or length between two points or two sides in a particular feature. A dimension line always has an arrow head at each end of the line, with a dimension centered adjacent to the line. Several ways to dimension civil design features are illustrated in Figure 7-12. Note that extension lines may be needed for dimensioning. Extension lines are used to extend the limits of two points or two sides far enough away from the feature so that dimensions can be added without crowding the drawing. It should be noted that an extension line should not touch the physical boundary of that feature; a small gap is typically used between the line and the boundary (Figure 7-12).

Leader

A leader is a line connecting a callout to the feature or portion of the feature shown. An arrowhead is used at the end of the leader, pointing to the feature. It is important that the

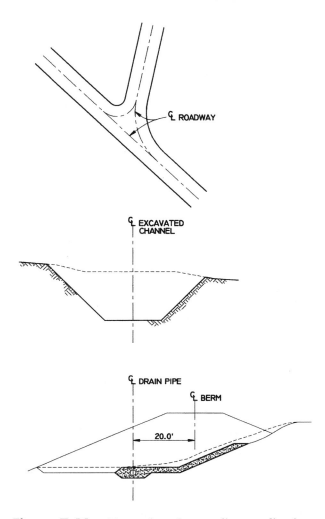

Figure 7-11.   Examples of centerline applications.

arrowhead touches the physical boundary of that feature. In addition, the leader should be connected to the beginning or end of a callout, but not connected at some intermediate location of that callout. There are several common ways to draw a leader, as shown in Figure 7-13, and all of them are quite acceptable. The important thing is to be consistent.

Hidden Line

A hidden line is used to show a physical boundary that is not directly visible or covered in a particular view, such as an internal opening or structure behind the section being shown. Examples where hidden lines are used include buried features, such as pipes, conduits, and footings. Hidden lines are also commonly used to show existing ground surfaces beyond the limits of excavation in a cross section. The use of hidden lines for a particular view should be minimized to improve readability and simplicity. When too many hidden lines are shown, the designer should consider another view to remove some of the hidden lines.

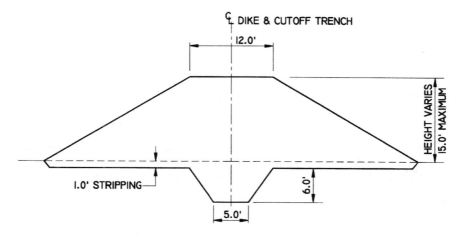

Figure 7-12.   Example of dimension lines and extension lines.

Phantom Line

A phantom line, as the name suggests, is used for something that is not in the view presented. For example, it can be used to show a feature that is projected onto a section from beyond that section. In other instances, it can be used to show an outline of a feature that is deliberately omitted in the view for clarity and simplicity, or it can be used to show the outline of a finished structure in a partially constructed stage to illustrate overall relationships. Figure 7-14 illustrates some common applications of a phantom line.

Short Break Line

A short break line is used to isolate a typical portion of a feature in lieu of showing the entire feature itself for clarity and simplicity. It can also be used to define an isolated "window" in an elevation view to show what is inside or behind that view. Examples of the application of the short break line are shown in Figure 7-15.

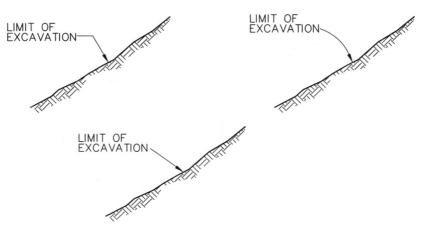

Figure 7-13.   Examples of drawing a leader.

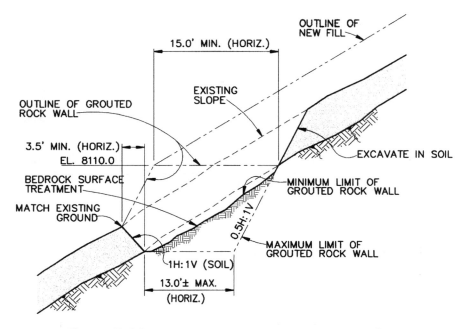

Figure 7-14.   Example of application of phantom line.

Long Break Line

A long break line is used to discontinue a long feature that would allow a very long plan view or sectional view to be shortened to a manageable size. A long break line is also used as a discontinued dimensional line for dimensioning features in a view in which the view itself has been broken by other long break lines. It is important that unique features of the view are not cut off between the long break lines, and that only repetitive information has been omitted. An example of the application of the long break line is shown in Figure 7-16.

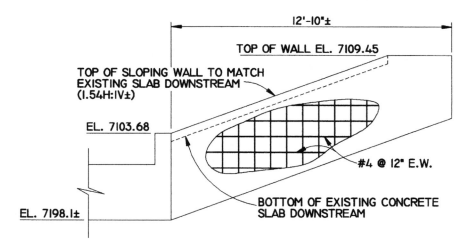

Figure 7-15.   Example of application of a short-break line.

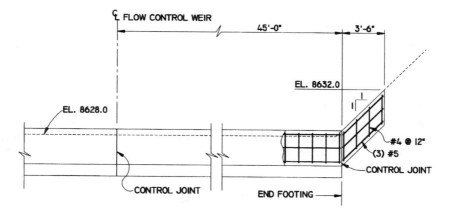

Figure 7-16. Example of application of a long-break line.

Match Line

A match line is an arbitrarily chosen splicing point for a view because the entire view itself does not fit within the length of the drawing sheet. The remainder of the view is normally continued on the same sheet, beginning at the match line, or on the next sheet. A match line allows the entire view of a very long feature to be shown without losing any part of that feature. A match line should always be identified with a number or other designation, including the sheet number where the remainder of the view is continued.

Contour Line

A contour line is a line of constant elevation. Topographic contour lines are used in a topographic map, which is the essential base map in a civil design project. Development of a topographic base map is discussed in Chapter 3, which also summarizes some basic rules involving contour lines. Contour lines should be labeled with the actual elevations adjacent to the lines. It is a good practice to use all digits of the elevation in the label instead of only using the last two digits. Not all contour lines need to be labeled. Typically, only major contours are labeled, not intermediate contours. For a map with 2-foot (0.6-m) contour intervals, for example, contours can be labeled every 10 feet (3.0 m), unless in very flat areas, in which case, it will be useful to label intermediate contours. Contour lines should be solid lines for both existing grades and final grades, except where the grades are changed. In that case, the new contours should be solid lines, and the existing contours under the new contours should be shown as hidden lines.

Water Line

A water line is used to show the limits of moving or stationary bodies of water, such as streams, lakes, or reservoirs. A single water line is sometimes used at the lowest point of the streambed, or a water line is used on each bank of the stream. It is always a good practice to add an arrow along the water line for a moving stream to show the direction of the flow, even though one can deduce the flow direction from the topographic contours. The water line for a stationary body of water, however, always coincides with the topographic contour of the water elevation, because the surface of the water is always horizontal.

7.8   Effective Use of Line Weights

Line weight is a term used to denote the thickness of the line. Different line types use different line weights to best bring out the graphical quality of a civil drawing and to show the important information. When a drawing uses the same line weight for different line types, the drawing becomes "flat," and some of the key features in the drawing may not stand out as desired by the designer. On the other hand, when proper line weights are utilized, the drawing has a certain three-dimensional quality and, along with effective hatching and symbols, allows certain key features to stand out and becomes more easily understood by the contractor. Figure 7-17 is a comparison of two drawings, one with a single line weight for all

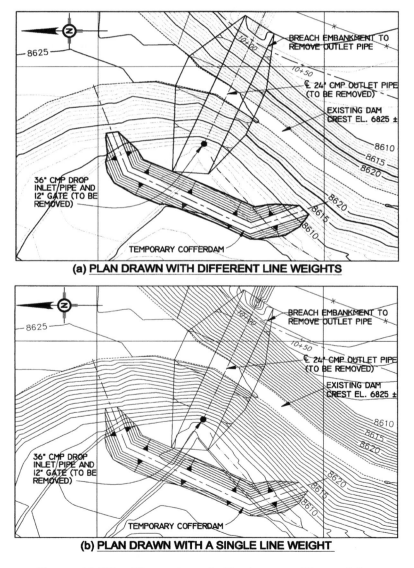

Figure 7-17.   Illustration of effective use of line weights.

Table 7-1. Lettering guidelines

Purpose	Font type	Size
Callouts, dimensioning, notes, abbreviations, scale	Simplex	0.10–0.12 inch
Headings for plan, section, detail, notes, legend, abbreviation	Arial	0.20 inch
Subheadings, Section or detail symbols	Arial	0.16 inch

of the lines, and the other with various line weights. This comparison illustrates the graphical and communication advantages of using different line weights.

Recommended line weights, or line widths, for the commonly encountered line types are shown in Figure 7-10. It is important to point out that these line widths represent guidelines only, and are not intended to be standards. Some designers may prefer using somewhat different line widths for various line types.

7.9   Lettering

Lettering is used in drawings for headings, callouts, dimensions, notes, and abbreviations. The traditional freehand lettering and hand-lettering guides are now replaced with electronic lettering fonts available in CAD software. There are literally hundreds of lettering

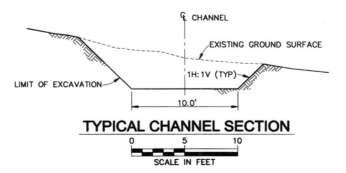

LEGEND
⊕   BENCHMARK

ABBREVIATIONS
E.F.   EACH FACE
TYP.   TYPICAL

NOTES
1. SEE SHEET 4 FOR GENERAL NOTES.
2. ALL EXCAVATION SLOPES SHALL CONFORM WITH OSHA REQUIREMENTS.

℄ CHANNEL

EXISTING GROUND SURFACE

1H:1V (TYP)

LIMIT OF EXCAVATION

10.0'

TYPICAL CHANNEL SECTION

0   5   10
SCALE IN FEET

Figure 7-18.   Illustration of proper use of lettering size.

fonts available, but surprisingly, only a few fonts (e.g., Simplex and Arial) are used in civil engineering drafting. In engineering drafting, fonts should be selected based on clarity and simplicity. Use of artistic fonts does not enhance or improve the technical quality.

The use of proper letter sizes cannot be overemphasized. The most common problem is undersized lettering. In general, lettering smaller than 0.10 in. (2.5 mm) should not be used. The use of a minimum lettering size is dictated, in most cases, by the fact that most drawings are now reproduced in half-size format, and the use of a minimum lettering size will assure that all of the lettering will be legible when reduced to half size. Most engineering firms and agencies maintain their own drafting standards, which dictate appropriate letter sizes. Table 7-1 provides a set of guidelines for lettering.

Figure 7-18 is an illustration of lettering using the guidelines from Table 7-1.

CHAPTER 8

LEGEND, ABBREVIATIONS, AND NOTES

8.1 Legend and Symbols

A legend is needed to define graphical symbols and graphical notations in drawings. Graphical symbols and notations should be defined on the sheet where they are first used, or on the general note sheet, and a note can be used on other sheets thereafter to refer to the legend. Not repeatedly defining symbols and notations on every sheet saves drafting time, and more importantly, avoids inconsistency in legends on the drawings when changes to the symbols or notations are required.

The following graphic symbols, hatchings, and notations should be adequately defined in the legends:

Symbols

Symbols are used to represent certain features on the drawings. These features may be boreholes, test pits, control points, vegetation, fences, roads, streams, pipes and culverts, and wetlands, all of which are commonly encountered in civil design work. There are many different symbols available to represent these features, and all of them are acceptable as long as they are used consistently within the drawing package and are properly defined in the legend. Figure 8-1 is a set of symbols that may serve as guidelines for designers who do not currently have their own standard symbols.

Hatching

Hatching is a two-dimensional pattern that is used to represent different materials. Materials used in civil engineering construction include different types of earthfill, bedrock, riprap, gravel and sand backfill, steel, concrete, masonry, wood, and wire mesh. Shading is also considered hatching. Figure 8-2 is a set of hatching patterns that may serve as guidelines for designers who do not currently have their own standard hatching patterns.

A special set of hatch patterns is commonly used by geotechnical engineers to represent subsurface conditions. These patterns are typically used in simplified borehole logs or test pit logs that may be included in construction drawings. Figure 8-3 is a set of hatch patterns of earth and rock materials that can be used to represent subsurface conditions.

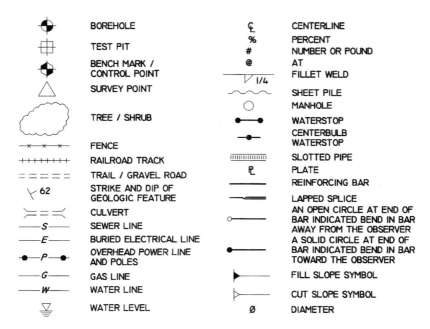

Figure 8-1. Guidelines for symbols.

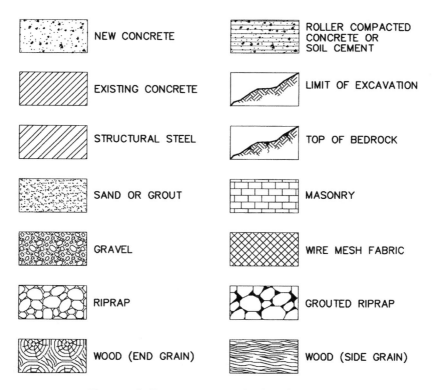

Figure 8-2. Guidelines for hatch patterns.

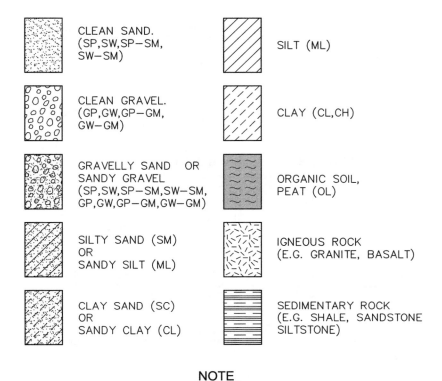

NOTE
SOIL DESIGNATIONS INSIDE PARENTHESES
ARE BASED ON UNIFIED SOIL
CLASSIFICATION SYSTEM.

Figure 8-3. Guidelines for hatch patterns for subsurface conditions.

Notations

Each set of design drawings contains a number of special notations that are unique for that project, depending on the imagination and creativity of the designer. For example, a set of notations can be used to call out the rows and columns of foundation piles or anchors. The legend should always contain a description of notations for section cuts and detail callouts. Different designers use different notations for section cuts and detail callouts, and all of them will accomplish the intent as long as these notations are defined in the legend and used consistently throughout the package. Figure 8-4 provides one method for section cuts and detail callouts. The advantage of this system is that, in the section or detail bubble, the following information is known:

- Section or detail designation is located on the top hemisphere.
- The sheet number where the section is cut or where the detail is referenced is shown on the left lower hemisphere.
- The sheet number where the section or detail is shown is shown on the right lower hemisphere.

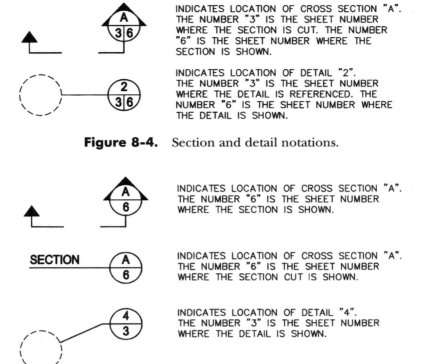

Figure 8-4. Section and detail notations.

Figure 8-5. Alternative notation for section and detail.

Some designers prefer to use a simpler system of section cuts and detail callouts, as shown on Figure 8-5. In that system, there is only one sheet number in the lower hemisphere of the section or detail bubble. In the section cut or detail callout, the sheet number refers to the sheet where the section or detail is shown. In the section or detail, the sheet number refers to the sheet where the section cut or detail callout is made.

8.2 Abbreviations

Abbreviations are used in drawings to shorten drawing callouts on plans, sections, and details, and to facilitate ease of reading the callouts. Abbreviations, once defined in the list of abbreviations on the general notes sheet, should be used consistently throughout the entire design. For example, *El.* or *Elev.* can be used for the term *elevation,* and the designer should use only one. In general, an excessive use of abbreviations in the drawing should be discouraged, and only the most common abbreviations should be used. In addition, any given callout should contain no more than one or two abbreviations. The following is an example of improper use of abbreviations:

½" CS PLATE, W/4-3" Φ HOLES DRILL THRU

APPROX.	APPROXIMATE	HR	HOUR	RCC	ROLLER COMPACTED
B.F.	BOTTOM FACE	HORIZ.	HORIZONTAL		CONCRETE
B.M.	BENCH MARK	I.D.	INSIDE DIAMETER	RCP	REINFORCED CONCRETE
B.O.B.	BOTTOM OF BOREHOLE	I.F.	INSIDE FACE		PIPE
B.O.T.P.	BOTTOM OF TEST PIT	INV.	INVERT	REINF.	REINFORCEMENT
C	CENTIGRADE	IN.	INCH	REQD.	REQUIRED
C.C.	CENTER TO CENTER	JT.	JOINT	REV.	REVISION
CFM	CUBIC FEET PER MINUTE	KIP	KILOPOUNDS	S	SOUTH
C.I.	CAST IRON	LB.	POUND	SCH.	SCHEDULE
C.J.	CONSTRUCTION JOINT	L.P.	LOW POINT	SH.	SHEET
CLR	CLEAR	LG.	LONG	S.S.	STAINLESS STEEL
CMP	CORRUGATED METAL PIPE	LONG.	LONGITUDINAL	SECT.	SECTION
CONC.	CONCRETE	MAX.	MAXIMUM	SIM.	SIMILAR
CONT.	CONTINUOUS	MIN.	MINIMUM	SQ.	SQUARE
CT.J.	CONTRACTION JOINT	M.H.	MANHOLE	SPEC	SPECIFICATIONS
DEG.	DEGREE	MFR.	MANUFACTURER	STA.	STATION
DIAG.	DIAGONAL	NGVD	NATIONAL GEODETIC	SYM.	SYMMETRICAL
DWL	DOWEL		VERTICAL DATUM	STD.	STANDARD
E	EAST, EASTING	N	NORTH, NORTHING	T&B	TOP AND BOTTOM
EA.	EACH	N.I.C.	NOT IN CONTRACT	T.F.	TOP FACE
E.F.	EACH FACE	N.T.S.	NOT TO SCALE	THK.	THICK
E.J.	EXPANSION JOINT	N.F.	NEAR FACE	TOPO.	TOPOGRAPHY
EL.	ELEVATION	O.C.	ON CENTERS	TYP.	TYPICAL
E.W.	EACH WAY	O.D.	OUTSIDE DIAMETER	T.O.W.	TOP OF WALL
EXIST.	EXISTING	O.F.	OUTSIDE FACE	U.N.O.	UNLESS NOTED
F	FAHRENHEIT	O.W.	OUTLET WORKS		OTHERWISE
F.F.	FAR FACE	P.C.	POINT OF CURVATURE	VERT.	VERTICAL
FIG.	FIGURE	P.T.	POINT OF TANGENCY	W	WEST
FT.	FEET	P.D.	PLAIN DOWEL	W/	WITH
FTG	FOOTING	P.I.	POINT OF INTERSECTION	W/O	WITHOUT
FIN.	FINISH	PL.	PLATE	W.S.	WATERSTOP
GA.	GAGE	PREFAB.	PREFABRICATED		
GAL	GALLON	PSI	POUNDS PER SQUARE INCH		
GALV.	GALVANIZED	PSF	POUNDS PER SQUARE FOOT		
GR.	GRADE	PVC	POLYVINYL-CHLORIDE		
GPM	GALLONS PER MINUTE	R	RADIUS		
H.P.	HIGH POINT, HORSE POWER				

Figure 8-6. List of abbreviations.

Instead, the above callout should be replaced with the following:

½" CARBON STEEL PLATE, WITH FOUR-3" DIA. HOLES DRILL THROUGH

Figure 8-6 is a list of commonly used abbreviations that are encountered in typical civil design drawings.

In general, abbreviations should not be used in the following places:

- On drawing or sheet titles.

- In notes.

- In definitions of abbreviations.

- In section or detail titles.

8.3 Notes

Notes are used on the drawings under the following circumstances:

- To provide additional clarification to a drawing detail.

- To supplement a drawing callout or an extension of a drawing callout.

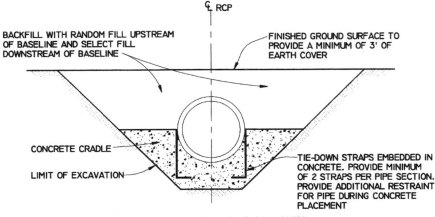

a. SECTION WITH LONG CALLOUTS

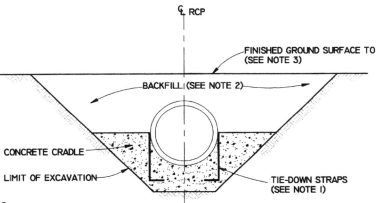

NOTES

1. STRAPS SHALL BE EMBEDDED IN CONCRETE CRADLE. PROVIDE A MINIMUM OF 2 STRAPS PER PIPE SECTION. PROVIDE ADDITIONAL RESTRAINT FOR PIPE DURING CONCRETE PLACEMENT.
2. BACKFILL WITH RANDOM FILL UPSTREAM OF BASELINE, AND WITH SELECT FILL DOWNSTREAM OF BASELINE.
3. PROVIDE A MINIMUM OF 3 FEET EARTH COVER ABOVE PIPE.

b. REVISED SECTION WITH SHORT CALLOUTS AND NOTES

Figure 8-7. Example of effective use of notes.

- To identify the basis or source of a design detail or data shown on the drawings.

- To refer to other sheets or specifications for the source of information, other notes, or other legends not contained on the current sheet.

- To make a statement regarding the contractor's duties and responsibilities that cannot be expressed graphically on the drawings but are directly related to information shown on the drawings or specifications.

- To define additional abbreviations not contained in the standard list of abbreviations.

- To suggest sequence of construction, caution, or other special requirement for a particular design feature or component.

Where possible, the preferred location for notes is at the lower right corner of the drawing sheet just above the title block. Notes should not be scattered all around the drawing, because it can be difficult to locate them and they are easily missed. Notes should be numbered consecutively for each drawing (1, 2, 3, etc.). Note numbers should not continue from one sheet to the next.

Identical notes should not be repeated on multiple sheets. Notes should be stated on the first sheet when they are needed, and referred to from other sheets.

Figure 8-7 is an example drawing showing how notes are used to make drawings easier to read and understand. Long callouts are shortened significantly by the use of notes.

CHAPTER 9

DRAWING PRODUCTION TECHNIQUES

9.1 General

This chapter contains a discussion of drawing production techniques for civil engineering design not covered in Chapters 6 through 8. These techniques can be used to improve efficiency, to enhance the graphical quality, and to clarify technical and construction requirements. The role and practicality of three-dimensional drawings in civil design are discussed. This chapter also contains the technique and minimum requirements for checking drawings during production.

9.2 Establishing Catch Points and Catch Lines

One of the most basic graphic techniques in civil engineering design is the establishment of catch lines for cut-and-fill construction. The role of catch lines in civil engineering design is to define cut-and-fill limits on the drawings. A catch point is the intersection of a cut or fill slope with the existing ground surface. A catch line is a line connecting the catch points. The determination of a catch line is based on descriptive geometry. *Descriptive geometry* is the science of graphic representation and the solution of spatial relationships of points, lines, and planes by means of projections (Giesecke 1975). Figure 9-1 illustrates the intersection of an excavation face (cut) with the existing ground surface, and Figure 9-2 illustrates the intersection of a fill slope (fill) with the existing ground surface.

Two examples are used to illustrate the establishment of catch lines. Figure 9-3 shows the establishment of excavation limits for a road cut. Figure 9-4 shows the establishment of fill limits for a detention embankment. A discussion of each example is given below.

Excavation Limits for Road Cut

A 30-foot (9.1-m)-wide roadway is required to pass through a topographic knob in bedrock. The knob will be removed to the roadway grade by excavation. For simplicity, the road grade across the knob is assumed to be flat at elevation 2160 feet. The permanent cut slope in the rock will be at 1H:1V (1horizontal:1vertical). Figure 9-3(a) shows the plan of the cut, and Figure 9-3(b) shows the cross section.

To determine the limits of excavation on the plan view, the centerline and the limits of the roadway are drawn first. From the edge of the road ditch (e.g., the north side), catch

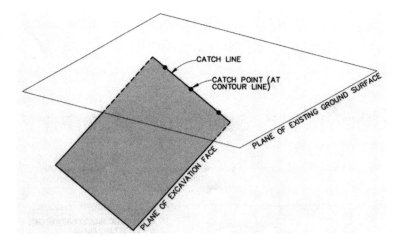

Figure 9-1. Schematic illustration of a catch line from excavation.

points for the excavation limits are established by calculating and then measuring the distance for each catch point from the edge of the road (at elevation 2160 feet) to meet a particular existing contour. Since the cut slope is at 1H:1V [that is, an offset distance of 5 feet (1.5 m) is needed for every 5-foot (1.5-m) rise in elevation], it would require a horizontal distance of 5 feet to establish the catch points to meet the 2165-contour; point a' and a'' are established this way. Similarly, points b' and b'' are established by offsetting a distance of 10 feet (3.0 m) from the edge of the road to meet the 2170-contour. The catch points thus established are connected to form a continuous limit of excavation on the north side of the road. The limit of excavation on the south side of the road is drawn using the same technique.

Fill Limits for Detention Embankment

A detention embankment dam is required to temporarily store storm water runoff on a stream. The embankment crest is at elevation 1348 feet, with a crest width of 20 feet (6.1 m).

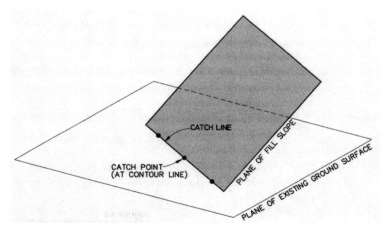

Figure 9-2. Schematic illustration of a catch line from filling.

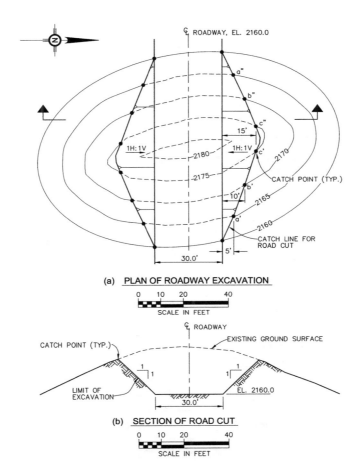

Figure 9-3. Example of catch line determination for road cut.

The embankment will have a 3H:1V upstream slope, and a 2H:1V downstream slope based on stability considerations. A 48-in. (1219-mm) culvert will be located at the bottom of the embankment to drain the detention pond. Figure 9-4(a) shows the plan of the embankment, and Figure 9-4(b) shows the cross section.

To determine the limits of the embankment on the plan view, the centerline and the limits of the embankment crest are first drawn. From the edge of the embankment crest (e.g., the east [downstream] side), catch points for the fill limits are established by calculating and then measuring the distance for each catch point from the edge of the embankment crest (at elevation 1348) to meet a particular existing contour. Because the fill slope on the downstream side is at 2H:1V [that is, an offset distance of 10 feet (3.0 m) is needed for every 5-foot (1.5-m) drop in elevation], it would require a horizontal distance of 6 feet (1.8 m) to establish the catch points to meet the 1345-contour (that is, two times the 3-foot (0.9-m) change in elevation); point a' and a'' are established this way. Similarly, points b' and b'' are established by offsetting a distance of 16 feet (4.9 m) from the edge of the embankment crest to meet the 1340-contour (that is, two times the 8-foot (2.4-m) change in elevation). The catch points thus established are connected to form a continuous toe of fill line on the east side of the embankment. The fill line on the west (upstream) side of the

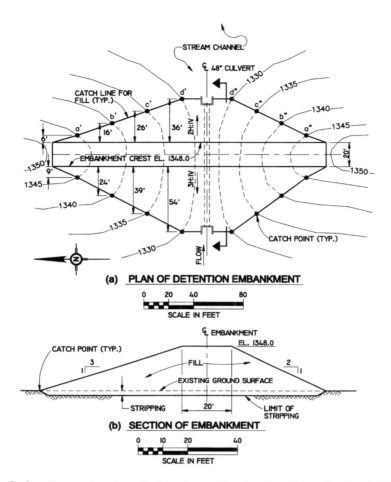

Figure 9-4. Example of catch line determination for detention embankment.

embankment is drawn using the same technique, except that the upstream slope has a 3H:1V slope (that is, an offset distance of 15 feet (4.6 m) is needed for every 5-foot (1.5-m) change in elevation). These catch points are connected to show the upstream heel of the embankment.

Before computer-aided drafting (CAD), establishing catch lines for civil design was a tedious undertaking, especially for complicated excavation or fill plans. With CAD, catch lines can be determined on the computer (see Chapter 11) with certain software. With or without the assistance of CAD, a civil designer must understand how cut-and-fill limits are established, at least in principle, and must be able to check CAD drawings for cut-and-fill accuracy.

9.3   Effective Use of Hatching and Shading

Hatching and shading are graphical representations of different material types. They are generally used to show separation among different materials, or to make certain features stand out. This graphical enhancement is necessary because construction drawings are not generally drawn in colors, and a black-and-white medium is used for original drawings and copies. A list of suggested hatching patterns is contained in Chapter 8. Shading can be considered as hatching, but shades of different darkness or lightness are used instead of graphical symbols.

Some guidelines are provided below for using hatching and shading in civil design drawings:

- A legend is required to define the hatching and shading used in the drawings. The only exception might be the standard concrete symbol, which is so universally used that a definition is generally not required. It is important to point out that the legend cannot replace a material description on the drawing detail itself.

- In addition to a legend, hatching or shading can also be called out in the drawing itself, using phrases such as "Existing Concrete (Shown Cross-Hatched)," or "Area to be Stripped (Shown Shaded)."

- Different hatched patterns are very effective when a detail contains several different materials that abut one another. The graphic symbols allow the various materials to stand out.

- Graphic hatching and shading should not be used when there is inadequate space in the drawing for them. The added symbols in a small drawing space would only complicate the drawing instead of achieving the intended purpose, especially when drawings are printed half size.

- Even when ample space is available in the drawing for hatching, the symbol does not have to fill in the entire area that it represents.

- Hatching and shading should not be used behind callouts or dimensions.

Figure 9-5 contains two design details in which hatching and shading are used effectively.

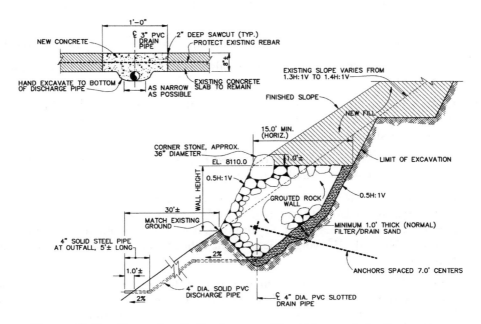

Figure 9-5. Example of effective use of hatching and shading pattern.

9.4   Use of Callouts and Dimensioning

Callouts are narrative descriptions of drawing components and features. A callout is used in conjunction with a leader and arrowhead pointing to the feature being described. Graphics and callouts on a drawing work hand in hand. Whereas the graphics shows where the various features and components are located, callouts define those features and components. Some general guidelines for the effective use of callouts are given below:

- Callouts should be as short as possible. When a long callout is required, an abbreviated callout should be used, with a reference to a note on the drawings (see Section 8.3). The main reason for this practice is to avoid using excessive space around a drawing detail for one particular callout; the space may be needed for other important callouts, or excessively long callouts may interfere with other sections and callouts in a drawing. For example, when a foundation subgrade is required, do not use a long callout such as "The contractor shall proof-roll the foundation subgrade by first scarifying the subgrade for at least 6 inches, moistening the subgrade, and compacting with at least two passes of an approved roller." Instead, a short callout should be used, such as "Proof-roll foundation subgrade (see Note 2)." The note being referenced should then contain the details of the requirement: "Note 2: Subgrade to be proof-rolled shall be scarified for at least 6 inches. Moisten the subgrade prior to compaction, and compact with at least two passes of an approved roller." If compaction requirements are already included in the specifications, then the note should be: "Note 2: Subgrade to be proof-rolled shall be in accordance with specifications."

- Avoid using abbreviations to the extent possible. Only the most commonly used abbreviations should be used, such as "EL. (ELEVATION)," "TYP. (TYPICAL)," "STA. (STATION)." All abbreviations should also be defined in the drawings (see Section 8.2).

- Use consistent terminology within the drawings and specifications (see Section 14.4).

- Avoid repeating requirements that are already in the technical specifications (see Section 14.4).

- Leaders for callouts should not cross each other to the extent possible. A callout should be located as close to the drawing feature as possible (i.e., the use of long leader lines should be minimized).

- Leaders for callouts should be connected to the beginning or the end of the callouts, but not at an intermediate point.

- In general, callouts should be located outside a drawing detail, and they should be fanned out in a balanced fashion around the drawing. In some cases, such as when ample space is available inside the drawing, it is quite acceptable to put the callout on the inside of the drawing, as long as the callout is outside the hatched or shaded patterns.

- When a drawing detail contains many repetitive features, it is not necessary to label every one of these features. Instead, only one of the typical features need to be labeled, with the abbreviation "TYP." in parentheses. This technique is effective in reducing the number of callouts, and thus simplifies the drawing. Figure 9-6 illus-

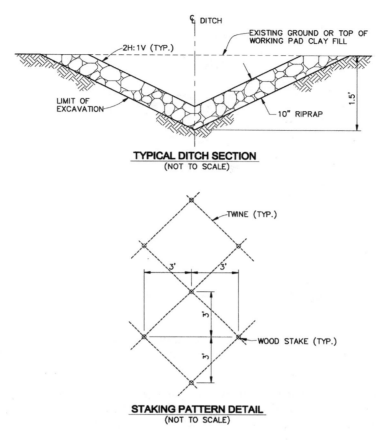

Figure 9-6. Using "typical" for callouts.

trates this technique. In one example, both ditch slopes are 2H:1V (2horizontal: 1vertical), and only one of the slopes needs to be labeled. In another example, the twine and wood stake shown on the repetitive staking pattern need to be labeled only once.

Dimensioning is the process of labeling linear distances or angles on the drawings. Dimensions are considered callouts and are accomplished with dimension lines, extension lines, and lettering. As such, discussion in Section 7.7 for dimension lines and extension lines, and discussion in this chapter for callouts, are applicable to dimensioning.

The following is a list of recommended guidelines on dimensioning civil design drawings:

- To conserve drawing space, only symbols or abbreviations should be used for the units involved. For example,

For 6.3 feet	use 6.3′
For 7 feet 4 inches	use 7′–4″
For 47 degrees	use 47°
For bearing	use, for example, N47°31′ 27″ E

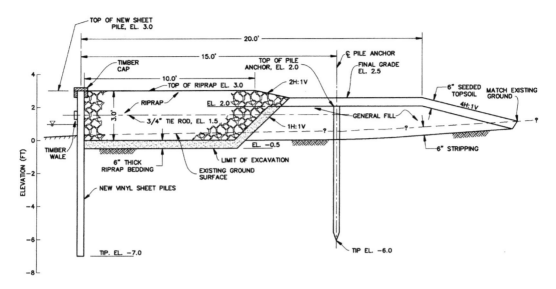

Figure 9-7. Example of stacking dimensions.

- To the extent possible, dimensions should be shown around the outside of the drawings. Dimensions placed inside a drawing are acceptable, provided that they are not placed directly over graphic symbols, hatching, or shading.

- When several dimensions are stacked on top of one another, the shortest dimensions should be located nearest the drawing, and the dimensions can progressively increase toward the top of the stack with an overall dimension on the outside. An example of this technique is shown in Figure 9-7.

- Dimensions that can be measured practically during construction should be shown on the drawings. For earthwork, the practical tolerance is plus or minus 0.1 foot (30 mm), and earthwork dimensions or elevations should be shown in decimals to the nearest 0.1 foot (30 mm). Concrete structures can be built and finished to the nearest ⅛ inch (3 mm), and concrete dimensions should be shown in feet and inches to the nearest ⅛ inch (3 mm). Fabricated metal structures are built more accurately than concrete structures, and dimensions can be shown in feet and inches to the nearest ¹⁄₃₂ inch (1 mm), and, in some cases, to the nearest micron. Field erection of concrete or fabricated metal structures, however, should be specified within an accuracy of 0.01 foot (3 mm) for controlled structure elevations.

9.5 Use of Scaled and Unscaled Details

All drawings can be categorized into three groups: scaled, exaggerated scaled, and unscaled. A discussion of each group of drawings is given below.

Scaled Drawings

In general, all drawings should be drawn to scale to the extent possible. When a drawing—whether it is a plan, profile, section, detail, or elevation—is drawn to scale, the first design decision by a designer is the selection of a proper scale (see Section 6.3 for scales for U.S. customary units, and Section 10.2 for metric units).

When a drawing is done to scale, the scale and the unit for the scale should be shown on the drawing. Recommendation on scale display is discussed in Section 6.4.

It is important to point out that, even when a drawing is done to scale, all pertinent controls and dimensions should be called out on the drawing. The controls and dimensions called out explicitly are the basis for construction. It is unreasonable to expect the contractor to scale the dimensions from a full-size or half-size drawing to obtain any missing information that is not shown. Because of the limited accuracy in hand measurements, hand measurements by the contractor may lead to misinterpretation. When information is missing from a scaled drawing, it is the contractor's responsibility to request the required information from the design engineer.

Exaggerated Scaled Drawing

A drawing is exaggerated when the vertical scale is different from the horizontal scale. In a long profile view, for example, the longitudinal dimensions usually far exceed the vertical dimensions, and it is customary to exaggerate the vertical dimensions to obtain a sufficient resolution for features along the profile. Other than in a long profile, it is recommended that the use of exaggerated scaled drawings be minimized. Exaggerated scaled drawings are somewhat awkward to draw, and the resulting graphics are contorted and unnatural. For example, in an exaggerated drawing, a cylindrical pipe will appear elliptical, a thin concrete footing will appear thick, and a flat slope will look steep. When exaggeration is used in a drawing, both the vertical and horizontal scale bars and units should be called out.

Care should be exercised in estimating quantities from an exaggerated scaled drawing using a planimeter. Some planimeters cannot measure quantities when the vertical and horizontal scales are different, but some of the new planimeters can do so by specifying both scales. Exaggerated scale drawings will not introduce a problem for quantity takeoffs on the computer.

Unscaled Drawings

Unscaled drawings should be used only for simple details or details where all of the relevant controls and dimensions are called out on other drawing details. A drawing not drawn to any scale should be labeled "Not to Scale" or "No Scale" under the title. The abbreviation NTS is also commonly used. Preparing not-to-scale drawings has the following incentive from a production standpoint:

- The drawing can be drawn quickly without attention to the actual dimensions.

- When part of the unscaled drawing is changed, only the dimensions need to be changed, and it may not be necessary to change the graphics portion, thus saving some drawing time.

An unscaled drawing can also be used to illustrate a construction method or sequence when the dimensions are not critical. Because a scale is not used, these drawings cannot be used to take off quantities for a cost estimate.

It is advisable that an unscaled drawing be drawn somewhat to scale to maintain the relative appearance of the design feature. Unscaled drawings should not be drawn like a cartoon, without regard for actual dimensions, sizes, and shapes of a design feature.

Several examples of details that are not drawn to scale are illustrated on Figure 9-8.

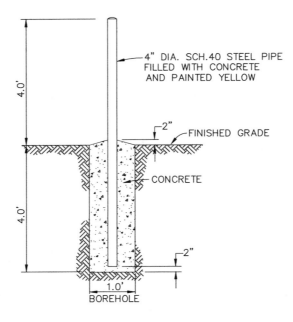

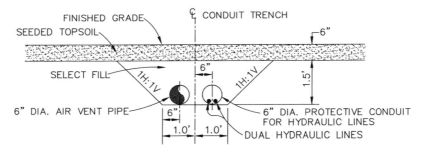

Figure 9-8.   Examples of details drawn not to scale.

9.6 Enlarging Details

Details show additional information on a particular feature of a design that cannot be conveniently shown because the scale of the overall feature is too small (see Section 7.6). Instead of calling out a detail, and then showing that detail separately on the same sheet or on a different sheet, a technique that is quite effective is to use an enlarged "bubble" acting essentially as a magnifying glass. The detail thus enlarged is still "attached" to the view, and generally, the enlarged scale does not need to be specified as long as the pertinent information or dimensions are called out. In other words, the enlarged detail shown can be drawn not to scale. It is important to note that this technique is effective when the detail to be enlarged will not be repeated elsewhere in other features of the design. If a detail is recurring, it is better to call out the detail with a specific designation (see Section 7.6), and show the detail separately.

Figure 9-9 shows an example of detail enlargement on the same drawing view.

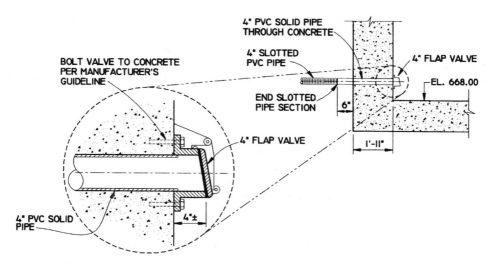

Figure 9-9. Enlarging details on same view.

9.7 Distinguishing New and Existing Work

This discussion applies to both new construction and modifications to existing facilities. For new construction, the distinction is between new grades and existing grades of the ground surface. In either case, it is important to distinguish on the drawings between existing and new facilities or features, and what is in the contract scope of work and what is not, and that distinction should be clear.

For new construction, the existing topographic contours under the new work should be shown as dashed lines (see examples in Figures 7-3, 9-3, and 9-4), and all new contours and existing contours that remain unchanged should be shown as solid lines. In some cases, leaving all of the existing contours within the new work would unnecessarily interfere with detailing the new work. In such situations, the practice is to dash in only the major contours, and delete the intermediate contours. It should be cautioned that, when this is done with CAD, intermediate contours that are deleted may be lost on that drawing file, and any future design changes to the new work might not include these contours. Therefore, it is recommended that dashing of the major contours and deleting of the intermediate contours be performed when the design is near completion, and additional major changes to the design are not anticipated.

In civil design, the most frequently encountered situation is work involving excavation and new fill. Excavation, whether it is temporary or permanent, is considered as new work, and the limit of excavation relative to the existing ground surface is important design and construction information. Other than stripping, excavation should be shown both on plan (referred to as excavation plan) and in cross sections. Excavations shown in cross sections should include the existing ground surface and limit of excavation. Showing the limit of excavation without showing the existing ground surface would be meaningless. Without the existing ground surface, the contractor does not know the depth of the excavation or where the excavation meets the ground surface. The limit of excavation shown on the drawings can also be used for measurement and payment by designating the limit of excavation as neat line for excavation payment purposes. Common drafting practice is to add hatching below the line to emphasize that limit of excavation. It should also be noted that an estimate of excavation quantity (Chapter 22) is impossible if an excavation cross section does not include the original ground surface.

When construction involves rehabilitating existing facilities or modifying existing facilities, the general guideline is to show the following:

- Limit of removal (demolition) of existing structure

- Limit of existing structure to remain (or protect)

- Interface between new work and existing work

- Points requiring field measurements to confirm actual locations

Figure 9-10 is a simple cross section illustrating this principle. An existing concrete structure is being lowered 2 feet (0.6 m) by saw-cutting, and the structure will be rebuilt with new concrete. The section shows the limit of concrete demolition, the existing concrete slab to remain, and the outline of the new concrete. To effectively show these three components (demolition, existing concrete to remain, and new concrete), cross hatching is used for the demolished concrete, and a phantom line is used to outline the new concrete. Other combinations of hatching, shading, and line types can be used, depending on the preference of the designer.

It is important to clearly call out what is existing work and what is new work. When details of the new work are shown, including how the new work interfaces with the existing structure, the existing structure or topography should be shown as background using lighter line weights, and the new work should be emphasized. The technique of using lighter line weights and lighter objects to de-emphasize the background is called *screening*. Prior to CAD, screening was a two-step process that involved photographic production of a lighter original, and the lighter original was then used for drafting. With CAD, screening is done electronically and rapidly, and many levels of screening (e.g., 95%, 90%, 85%, 80%, etc.) can be accomplished. Examples of screened lines are shown on Figures 7-1 and 7-17.

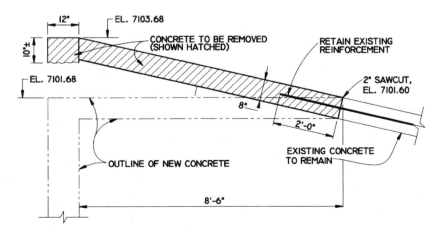

Figure 9-10. Example of distinguishing new work from existing facility.

9.8   Representing Symmetry

Many civil engineering structures are symmetrical in geometry and layout, with features identical in every respect on each side of the centerline. For a symmetrical structure, the entire structure is still drawn in its entirety on plan, except only half of the structure would need to be detailed. Callouts that are used to detail half of the structure but are intended for the entire structure should include the word *typical,* or the phrase *symmetrical about centerline,* which would suggest features on the other side of the centerline are similar (see Section 9.4 for use of callouts).

Strictly speaking, it is only necessary to draw half of a symmetrical structure or section because the halves are identical. There may be some savings in drafting time when drawings are prepared by hand. With CAD, when half of a symmetrical structure is drawn, the other half can be reproduced quickly by using the technique of mirroring (see Section 11.2). Therefore, showing only half of a symmetrical structure is hardly justified from a CAD production standpoint, and is not recommended.

There is one technique that can be used for preparing cross sections for a symmetrical structure (Figure 9-11). For simplicity, no dimensions or scales are used in this example.

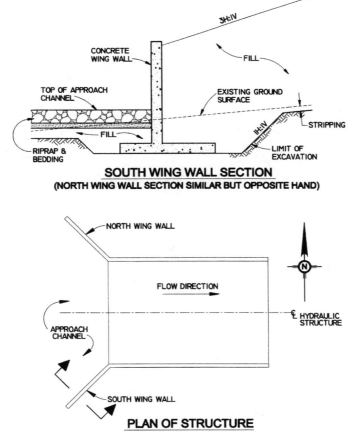

Figure 9-11.   Example of detailing symmetrical structures.

When one side of a symmetrical structure is shown in cross section, there is no need to show the cross section from the other side of the centerline. Instead, below the title for the cross section, a statement can be added to indicate that the other section is similar but opposite hand. For the south wing wall shown in the example, the phrase "North Wing Wall Similar But Opposite Hand" is sufficient to show the contractor that the same cross section and details can be used for both wing walls.

9.9   Use of Three-Dimensional Graphics

In general, three-dimensional graphical techniques, such as isometric, prospective, or oblique views, are not used in civil engineering construction drawings. The principal views, namely plan, section, and elevation, drawn in two dimensions, are still the basic graphical representations of civil design features. There are some instances in which a three-dimensional view may be more economical to produce than using the three principal views. Figure 9-12, which shows an intricate connection of a network of foundation drainpipes, is such an example. To show the same connection in two dimensions, at least three views (one plan view and two elevation views) are required.

In general, however, the use of three-dimensional views in civil construction drawings is not recommended. Three-dimensional views can be used for minor design features where multiple, two-dimensional views are more time-consuming to produce, and to provide clarity to the contractor for construction layout.

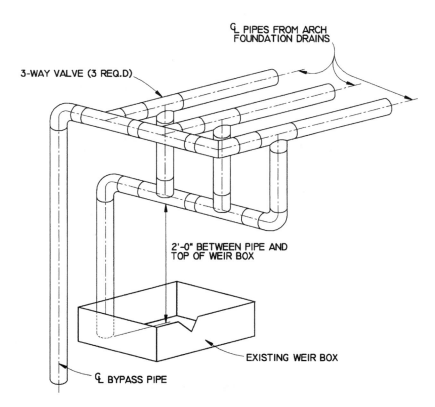

Figure 9-12.   Example of three-dimensional view.

9.10   Checking Drawings

Checking is part of the process of drawing production. This effort represents one of the most significant quality control protocols in applied engineering design, and therefore should be performed with the utmost care and precision. Significant errors may remain undetected if the drawings are not carefully checked and back-checked during the production process. The marking of progress drawings during production is also known as *red-lining* because a red pen is usually used for making changes on the drawings. Drawings that are not checked systematically by experienced design personnel undoubtedly will contain errors, regardless of how competent the designer is. It is also quite safe to say that drawings that are produced in a hurry to meet tight deadlines will have a high probability of errors, resulting in construction problems and claims.

There are several levels of checking that should be performed during the production of design drawings:

Checking for Drafting Errors

This first line of defense against errors in design drawing production should be performed by the design engineer. When the design drawing is first sketched by hand and submitted to the CAD drafter to be drawn electronically, the design engineer should check the CAD plot against the original hand drawing for accuracy. When the CAD drafter produces a drawing based on verbal or written instructions of design requirements from the design engineer, the design engineer should check the CAD plot against the instructions given. In addition, the design engineer should also check to make sure all of the views drawn and plotted are consistent with the instructions given. Specifically, the design engineer should check for the following items:

- Control points, baselines, and centerlines are drawn correctly. Survey and control information should be checked independently by the engineer.

- Elevations, distances, dimensions, and angles are represented correctly.

- The spatial relationships among project features are portrayed correctly in the views.

- Catch points and catch lines are drawn correctly.

- Drafting standards (line types, line weights, symbols, etc.) are applied correctly.

As discussed in Chapter 11, hand checking certain features of CAD drawings (such as northings and eastings of control points, stations, dimensions, and angles) may not be possible because CAD is significantly more precise than hand checking. For example, control points identified on CAD and expressed in 0.01-foot (3-mm) accuracy cannot possibly be measured and checked by the engineer, unless the engineer can do so independently on the computer. For those design engineers that are not trained in CAD, it is suggested that this type of information be checked independently by another CAD drafter. It is the responsibility of the design engineer to identify what part of a CAD drawing should be checked independently by another CAD drafter on the computer.

Checking for Technical Accuracy

Someone other than the designer should check for technical accuracy and constructability. This is normally done by a senior design professional in the organization, such as the project manager or principal-in-charge. It is sometimes performed by an outside consultant during an independent review. The responsibilities of technical design review are not to focus on minor drawing standards such as spelling errors or wrong graphic symbols or line types. Rather, the reviewer should focus on the following aspects of the design:

- Does the design meet the project scope of work?

- Are the design features and construction methods appropriate for the site conditions?

- Are the designs constructable and satisfactory according to health and safety requirements?

- How risky are the designs, from the standpoint of construction safety and safety of the structures?

- How many unknowns are present in the design?

- What is the potential for claims and changes during construction?

- Are there significantly more economical details for construction?

In order to meaningfully address these issues, the reviewer should perform a simultaneous review of the technical specifications, design report, bid schedule, and the engineer's cost estimate.

CHAPTER 10

Designing with the Metric System

10.1   General

Civil design discussed thus far has been based on the U.S. customary system of units. Despite numerous efforts in the private sector and in state and federal government to change to SI (System International) units (the metric system), the U.S. customary system of units continues to be the prominent system used in the United States. The American Society of Civil Engineers, for example, requires that all of their publications, such as journals and conference proceedings, be written in SI units, with conversion units allowed. Most of the state and federal departments of transportation have changed their specifications for bridge and road construction to the metric system, even though the U.S. customary unit versions are also available in some states.

It is not the purpose of this book to take a position as to which system is better. The entire book is written with U.S. customary units because it is still the dominant system used in the design and construction industry in the United States. That is also the system to which the author is accustomed. Nevertheless, the author has worked on several international design projects for which the metric system was required. This chapter is written for those designers that are interested in working in SI units.

In this chapter, design practices that are unique to the metric system are discussed and compared with the principles presented in this book. To illustrate the differences between the U.S. customary and metric systems, examples of drawings prepared with the metric system are shown.

10.2   Metric System Design Practice

The general principles to prepare design drawings are no different from those discussed in Chapters 5 through 9. All graphical techniques are the same for both systems. Some of the design features that are unique to the metric system are listed and discussed below.

Sheet Sizes

Full-sized metric sheets are either 817 mm × 570 mm, or 817 mm × 1105 mm. In the United States, even when a design is based on metric units, the drawing sheet sizes are

83

still in U.S. customary units because most reprographic copiers only use U.S. customary-size paper (see Section 5.1). Therefore, metric-sized sheets are rarely used in the United States.

Common Units

The common units and their abbreviations used in design drawings are:

Distance (civil drawings):	meter (m)
Civil drawing dimensions:	meter (m)
Structural/mechanical dimensions:	millimeters (mm)
Angular dimensions:	degrees, minutes, and seconds
Area:	square meters (m²)
Volume:	cubic meters (m³)
Temperature:	degree centigrade (°C)

Perhaps the most interesting practice is the use of millimeters for structural and mechanical dimensions. For comparison, in the U.S. customary system, the units of feet and inches are used for structural dimensions. In the metric system, structural dimension is expressed in millimeters to the nearest millimeter. Fractions of millimeters are not used because they cannot be practically measured during construction. For example, a structural dimension of 3 meters should be labeled 3000 mm on the drawing. However, the length of 3 meters can be labeled for civil drawings, such as depth of cut and fill, similar to the practice for U.S. customary units. The level of accuracy for earthwork construction is 0.03 m (3 centimeters).

Figure 10-1 illustrates a cross section of a civil drawing, and Figure 10-2 illustrates a structural cross section, both prepared using metric units.

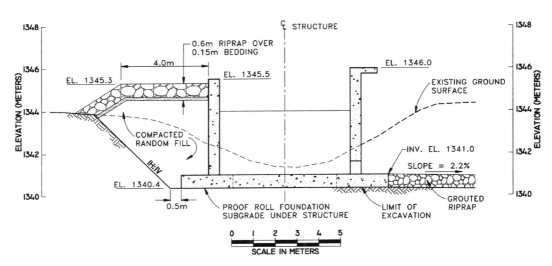

Figure 10-1. Example of a cross section prepared with metric units.

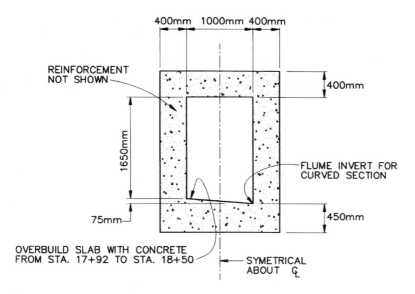

Figure 10-2. Example of a structure cross section prepared with metric units.

Metric Scales

Similar to the U.S. customary system, two sets of scales are available for metric drawings. The metric civil engineer's scale consists of the following common scales:

1:500 (1 cm = 5 meters)
1:1000 (1 cm = 10 meters)
1:1250 (1 cm = 12.5 meters)
1:1500 (1 cm = 15 meters)
1:2000 (1 cm = 20 meters)
1:2500 (1 cm = 25 meters)

The architect's scale that is used in structural and mechanical design consists of the following common scales:

1:100 (1 cm = 1 meter)
1:200 (1 cm = 2 meters)
1:250 (1 cm = 2.5 meters)
1:300 (1 cm = 3 meters)
1:400 (1 cm = 4 meters)
1:500 (1 cm = 5 meters)

10.3   Equipment and Products

The specification for manufactured equipment and products when using the metric system requires some discussion. Although most equipment and products manufactured outside of the United States are based on metric units, those manufactured in the United States are

Table 10-1. Reinforcing bar designations

U.S. customary units		Metric units	
Designation	**Diameter (in.)**	**Designation**	**Diameter (mm)**
No. 3	⅜	No. 10	10
No. 4	½	No. 13	13
No. 5	⅝	No. 16	16
No. 6	¾	No. 19	19
No. 7	⅞	No. 22	22
No. 8	1.0	No. 25	25
No. 9	1⅛	No. 29	29
No. 10	1¼	No. 32	32
No. 11	1⅜	No. 36	36
No. 12*	1½	No. 38	38
No. 14	1¾	No. 43	43
No. 16*	2.0	No. 51	51
No. 18*	2¼	No. 57	57

*Not ASTM standard bar size

still manufactured using U.S. customary units. When a domestic project, that is in the United States, is designed with metric units, the designer can still call out a certain pipe, gate, valve, or structural steel member in U.S. customary units, and the contractor can procure these products. For projects outside of the United States, the designer is faced with the following decision: Should he or she specify an American-made product or equipment (made with U.S. customary units) that needs to be imported, or should he or she specify a metric product that is available locally? The answer is usually determined by economics.

Because of frequent inavailability, one cannot simply convert a product or equipment from U.S. customary units to metric units in selecting a metric product or equipment in specifications. For example, a 42-inch (1067 mm) steel pipe may only be available in 1000 mm, and a ¼-inch (6.35 mm) steel plate may only be available in 6 mm stock. In the United States, reinforcing bars are numbered based on multiples of ⅛-inch; in the metric system, the reinforcing bars are numbered based on the nearest millimeters. For example, a No. 3 bar (⅜-inch) in the United States is equivalent to a No. 10 bar (10 mm) in metric, and a No. 7 bar (⅞-inch) in the United States is equivalent to a No. 22 (22 mm) bar in metric. Table 10-1 contains reinforcing bar designations in both U.S. customary units and metric units.

When combining metric equipment with U.S. customary unit structural or mechanical work, the designer is solely responsible for compatibility and clearances. Therefore, a designer would need to be knowledgeable in the standards and technical specifications regulating the manufacture of metric products and materials.

CHAPTER 11

COMPUTER-AIDED DRAFTING

11.1 Current Trend

In the United States, essentially all design drawings are now prepared with computer-aided drafting (CAD) software. The use of CAD software in engineering design became popular in the 1980s, when high-speed, affordable personal computers with large storage memories became available. The days of the drafting table, Leroy lettering tools, and rapidographs are all but over, but a good design engineer does not necessarily have to be a computer wizard or an expert in CAD software. Production of design drawings should be the responsibility of CAD drafters, who are technicians specifically trained for this type of work; however, there is a new breed of engineer who have CAD training in their education curriculum, and they are using the computer essentially as a design tool while also doing CAD production. Because the hourly rate for an engineer is significantly higher than that of a CAD technician, such a practice may not be cost-effective from a business perspective.

This chapter describes some common computer-aided design tools and their usefulness in civil engineering design, the advantages of CAD drawings, and the roles and responsibilities of design engineers and CAD drafters in the production of construction drawings. This chapter also contains some guidelines regarding the use and transmission of electronic design information, including design drawings, shop drawings, and design changes.

11.2 Computer-Aided Tools and Capabilities

An exhaustive list of CAD software packages is made impractical by the immeasurably high number of CAD software packages available, the rapid replacement of software packages because of the evolution of capabilities, and frequent product name changes caused by corporate buyouts and takeovers. This section, therefore, identifies and discusses—in terms of their general capabilities as a group—several common software programs used in the United States for civil engineering design.

In addition to the cost of constantly updating CAD software packages, there are two practical problems that should be considered by persons responsible for CAD production work:

1. **Archiving and retrieving drawings.** When a CAD program is updated or changed, the new version of the software may cause problems with file compatibility, preventing the recovery of archived files. To minimize the loss of electronic drawing

data, users should keep old programs active for a period of time before discarding them. A holding period of 10 years is suggested.

2. **Continuing education of the production staff.** When a new or updated program is used, the production staff should be trained accordingly. Although the business expenses associated with this type of continuing education are necessary, they may be difficult to sustain because of the frequency with which new and updated versions are released. Without proper training, however, productivity can fall while the staff constantly struggles with a steep learning curve. Therefore, from a business production standpoint, the same program should be used for 3 to 5 years before it is replaced.

The following is a general description of some capabilities of common CAD software programs [e.g., AUTOCAD (Autodesk, Inc.; San Rafael, Calif.), MICROSTATION (Bentley Systems Inc.; Exton, Penn.), EAGLEPOINT (Eagle Point, Dubuque, Iowa)]. No attempt will be made to distinguish the capabilities of these programs from one another.

Drawing capability. From a graphical standpoint, all of these programs can generate a set of civil design drawings that are equal to or better than the hand-drawn drawings of the past. Some creative and artistic aspects of hand-drawn drawings are gone, and the creation of design drawings is becoming more technical and machine-oriented. Nevertheless, good-quality drawings are created by capable practitioners, and poor-quality drawings are created by less capable and inexperienced drafters. All CAD software programs contain an abundance of symbols, shadings, font types, and font sizes to represent all of the basic graphical requirements for civil design. For example, one excellent graphical capability in CAD is the ability to screen any line and object to enhance the graphic quality (see Section 9.7). Before the advent of CAD, the use of screening in drawings required an extra step in the screening process using the photographic method. CAD allows a nearly infinite variety of screening and shading without resorting to that extra step.

Computational capability. CAD software is no longer just for creating drawings. Essentially all CAD software has computational capabilities that previously required a designer. A good example is the survey computations that are routine in civil design. Vertical curve or horizontal curve computations can be performed rapidly as part of the CAD design process, and the CAD drafter now can easily identify points of curvature, points of tangency, etc., for curve design.

Improved accuracy. CAD software is a far more accurate design tool than the tools that were used for hand-drafting. For example, with an engineer's scale of 1 inch = 10 feet, lengths and distances in drawings can be estimated only to an accuracy of ± 0.1 foot. A protractor can only read an angle to within half a degree. On the other hand, the accuracy of CAD software is not dependent on the drawing scale, and lengths and distances can be measured to as many significant figures as are required. Using a computer, distances can be measured and identified to the nearest 0.001 feet, and angles can be measured and identified to the nearest second. For all practical purposes, the accuracy that can be physically surveyed or measured in the field will be adequate. In this sense, CAD drawings are considered very precise.

Drawing layout. Some CAD programs, such as EAGLEPOINT or the Land Development module of AUTOCAD, have the capability to layout plan, sections, and profiles for civil design. When the triangular irregular network (TIN) of an electronic base map is established, the CAD drafter can perform—in a fraction of the previously required time—numerous types of design work traditionally performed by hand by an engineer. These design tasks include establishing catch points for cut-and-fill construction and cutting cross sections and profiles.

Most CAD programs can also perform the following maneuvers, which are very time-consuming when performed by hand:

- Changing drawing scale—CAD drafting is performed on a 1:1 scale, and the drawing can be plotted at any desirable scale. This is especially useful when different scales are needed for a base map to show different levels of detail. Before CAD, changing the scale of a base map required photographic or xerographic enlargement or reduction, which is time-consuming and causes distortion of the map. With CAD, this process can be done electronically and quickly, and there is no distortion associated with enlargement or reduction of a drawing or map.

- Cut-and-paste work—There are many situations in which cutting and pasting or copying and pasting design layouts or details are required. CAD software can perform this type of work very quickly. For example, a standard detail can be modified quickly for a particular project, without requiring the entire detail to be drawn from the start.

- Mirroring—Mirroring is the technique of reversing a drawing or detail. When a detail is symmetrical, only half the detail needs to be drafted; the other half can be reproduced by the mirroring technique. Sometimes, an opposite view may be more desirable; and, instead of redrafting the entire detail, the opposite view can be produced using mirroring.

Quantity takeoff. Estimating quantities from design drawings is part of the design process and can be time-consuming, depending on the complexity of the design. Without CAD, the estimations (see Chapter 22) can be established using a scale or a planimeter, supplemented by additional cross sections or profiles where necessary. Using CAD software, this tedious process can be performed in less time and with significantly increased accuracy.

From a business standpoint, the use of a computer to produce design drawings offers great advantages in cost savings and production efficiency. When one compares the time required to produce a set of drawings the first time with and without CAD, its resulting cost savings are debatable. However, there is no question that changing drawings using CAD saves a significant amount of time. To address internal review comments, to address review comments by clients and regulatory agencies, and to update changes that are made during construction, design drawings are frequently changed during the various stages of design. The old method of making changes—erasing and redrafting by hand—is simply too time-consuming, and frequently an entire drawing requiring significant changes would need to be redrawn completely.

11.3 Roles and Responsibilities

CAD software is a design tool, and should be used as such. CAD software does not replace the technical training, experience, and judgment of an engineer. Regardless of the tool available for design, there should be a distinction between engineering design and engineering drafting. Before the advent of CAD, seasoned and experienced engineers performed most engineering design. Design responsibilities in some firms and agencies are now falling into the hands of CAD operators who are proficient in CAD software and computer skills, but have no engineering training or construction experience. Such a trend is indicative of the belief that all-powerful drafting tools, such as some of the sophisticated CAD programs, can replace the technical skills and judgment of an experienced engineer. In a recent special report in *Engineering News Record* (ENR 2000), building contractors voiced their opinions on the decline in quality of design in the following excerpts:

"The largest problem in the industry is poor contract drawing."

"Architects and/or engineers are not creating a product that contains all the necessary information. To save money, they use inexperienced personnel."

"Dimensions are a major problem."

"Drawings are getting worse."

"We see structural drawings that don't even have the grid dimensions."

"People come out of school who are fluent with a computer as if it's a video game... But if you analyze it, you realize the operators are only drawing lines on paper – not walls, floors and columns."

The graphical quality of design drawings has improved with the use of CAD software, but attractive drawings alone are inadequate for successful construction. Some of the problems cited above are not caused by the misuse of CAD; rather, they are caused by the management responsible for staffing and managing some design projects. The remaining portion of this section describes the roles and separate responsibilities of a design engineer and a drafter in the production of drawings using CAD software.

Role and responsibility of a design engineer:

- The design engineer should have primary and overall responsibility for the technical accuracy of the design within his or her area of expertise.

- The design engineer should be responsible for the constructability, health, and public safety of his or her design. A CAD drafter may not know that it is unsafe to design a very steep cut slope for a cohesionless soil, nor is he or she expected to know that a 1H:1V slope is too steep to support a riprap rock lining for a stream channel.

- The design engineer should document the design criteria and technical basis of his or her design. A CAD drafter can draw the detail of a reinforced concrete pipe or the detail of a post-tensioned anchor. The technical basis for using that type of pipe or that anchor size and corrosion protection should originate from the design engineer, rather than from the CAD drafter who can modify these details from previous projects but does not know why they are used in a certain way.

- The design engineer should be responsible for selecting the products and materials.

- So that adequate information is provided for the contractor, the design engineer should be responsible for selecting the geometric layout and dimensions of his or her design. Whether the engineer performed the graphic construction of his or her design by hand or using a computer, or simply give a set of instructions to the CAD drafter, is not important. What is important is that the decision on a particular aspect of the design originates from the engineer.

- The design engineer—or, occasionally, other staff working under his or her supervision—should be responsible for checking and back-checking drawings.

- When a CAD drafter provides quantity calculation data to the engineer, it is the responsibility of the engineer to check the data for accuracy and completeness before using them for cost estimating and bidding. Unchecked quantity calculations generated by the computer may lead to significant differences in construction paid quantities, resulting in claims and project-cost overruns.

Role and responsibility of a CAD drafter:

- The main role of a CAD drafter is to assist the design engineer in the preparation and production of design drawings. His or her main responsibility is to ensure that the drawings are prepared accurately based on instructions given by the design engineer.

- A CAD drafter should refrain from activities that exceed the bounds of his or her knowledge, such as introducing technical elements in the design drawings. When the design engineer specifies the centerline alignment of a trench excavation but neglects to include the side slopes of the excavation, it is the drafter's responsibility to request this missing information from the designer, and not unilaterally assume a slope that may not be safe, or may not comply with Occupational Safety and Health Administration (OSHA) requirements.

- A CAD drafter should be responsible for all aspects of the product he or she is producing—the accuracy and graphical aspects of the drawings, the plot, and the filing system. The drafter should have his or her own quality-control procedures to make sure all numbers (e.g., coordinates, stations, dimensions) on the drawings are entered correctly. The drafter should know what line types and line weights are appropriate for a given design feature. A CAD drafter is also responsible for using standard symbols, standard details, and industry or company standards for drawings.

- The CAD drafter has evolved into more than just a tracer. With a powerful computer and CAD software, he or she can perform a variety of chores that greatly reduce the involvement of the design engineer in the production of design drawings. Design engineers that fail to recognize opportunities such as these are not taking full advantage of modern technology. A CAD drafter can assist the engineer in laying out the plan, sections, and profile if the engineer provides him or her with the necessary survey control and other design and site information. A CAD drafter can identify dimensional data on the computer much more accurately than an engineer can with hand scales and tools. A CAD drafter can provide quantity data from the drawings for cost-estimating purposes.

11.4 Handling of Files

The ability to digitize design drawings and save the information as electronic files introduces a whole new set of issues on storage, transmittal, and professional liability. The old method of storing reproducible drawings in a flat file cabinet has given way to electronic media. Instead of hard copies, electronic files of design drawings, shop drawings, design submittals, and even bid documents are often transmitted. Sometimes, contractors and regulators ask for these electronic files from the designer for their use, and it becomes important to look into the issue of protecting the ownership and professional liability of the design engineer.

Storage of CAD Data

The development of large hard drives and efficient electronic storage media has provided engineers with the ability to store design drawings efficiently. In approximately the last 20 years, the size of hard drives on personal computers has increased from a few megabytes to many gigabytes. The size of storage disks has increased from less than one megabyte (diskette) to more than 100 megabytes (zip disk). The recent use of rewritable compact discs (CDs) allows up to approximately 700 megabytes of data to be stored on a single CD. At this time, the most reliable permanent storage medium for CAD data is on a compact disc. The permanent electronic copy should be stored in a separate site and in a fireproof cabinet. It is recommended that every design project be archived not only electronically, but also with a hard copy of reproducible vellum or Mylar.

Transmittal of CAD Data

Electronic files of CAD design are transmitted to clients for progress review and as final submittal. During bidding and construction, sophisticated contractors also use CAD to prepare their bids, shop drawings, and record drawings. Frequently, bidders and contractors request a copy of the electronic file to be sent to them for their use. Engineers also receive electronic files from contractors for shop drawing submittals and record drawings. E-mail and file transfer protocol (ftp) sites make transmittal of electronic data almost instantaneous. From a design engineer's point of view, numerous important questions and concerns exist regarding this free flow of design information:

- How does a design engineer protect his or her ownership and intellectual proprietary rights?

- How does a design engineer prevent others from altering his or her design drawings?

- How does a design engineer maintain a defensible paper trail for electronic CAD data sent and received?

- How does a design engineer prevent his or her CAD files from being corrupted or infected with electronic viruses?

There are no clear-cut answers to these questions. Some general guidelines are:

- All drawings sent electronically should have a hard-copy backup in the project file.

- Transmitting DWG files is not recommended. If only a plot is needed, then a plot file (PLT format) or a PDF file can be used. Files with PLT or PDF extensions cannot be altered.

- Electronic shop drawings received from contractors should not be altered, and should only be used to obtain a plot for review and comment. Review comments on shop drawings should be written in red on hard copies, and transmitted back to the contractor and owner.

- Electronic files must be routinely checked for compatibility with new software programs.

CHAPTER 12

CERTIFYING CONSTRUCTION DRAWINGS

12.1 Common Practice of Drawing Certification

When a set of construction drawings is completed, the common practice is for the engineer of record to certify the drawings. Each state has different requirements for certification, but in general, certification of a design by a professional engineer implies the following:

- The design is prepared by the engineer, or it is prepared under the control and direct supervision of that engineer.

- The engineer assumes full responsibility and liability for his or her portion of the design.

- The design is safe for public health, property, and welfare.

The certification process consists of stamping (or sealing), signing, and dating the drawings. Engineers routinely sign and stamp documents, but records indicate that these tasks are frequently mishandled and that they are the topic of most frequently asked questions for states' boards of registration for professional engineers. The information that must be disclosed in the certification process varies from state to state, but almost all of them require the engineer to stamp, sign, and date the drawings. In some states, such as Illinois, Washington, and Oregon, the engineer is also required to indicate, below the seal, the expiration date of the professional engineer registration.

Only final drawings and record drawings (formerly referred to as *as-built drawings*) must be certified before submission to the client. Draft drawings or drawings that are issued for review purposes do not require certification. In fact, certifying drawings that are not finalized can convey the false impression that the drawings are complete. Most states require that draft drawings or review drawings indicate that they are preliminary and not suitable for construction. Similarly, design drawings for a study, a feasibility design, or a conceptual design do not require certification. Therefore, a large callout, "PRELIMINARY—NOT FOR CONSTRUCTION," is usually added to those drawings.

Shop drawings requiring professional services or work should require certification by the engineer who prepares them. The State of Arizona (1991) provided the following criteria for situations that do not require stamping of shop drawings:

- Sizing and dimensioning information for fabrication purposes

- Construction techniques or sequences

- Components with previous approvals or designed by the engineer of record

- Modifications to existing installations that do not affect the original design parameters and do not require additional computations

All state boards of registration require engineers to certify their original drawings and to keep duplicate sets of those certified drawings. In the event that changes are made to an electronic version of drawings submitted to a client, the original set in the engineer's possession will be used for comparison. Most states allow copies to be made from the certified originals, and wet-ink signatures on the copies are not required if the originals are stamped and wet-ink signed and are in the possession of the engineer of record.

Because policies vary from state to state, the reader should check the requirements of his or her state to ensure strict adherence to state requirements.

12.2　Who Should Certify Drawings?

In general, the engineer responsible for a design should certify its drawings. For the following reasons, however, the engineer in charge might not be easily identifiable in some cases:

- Design projects, whether in the public or private sector, are managed by project managers, who may only be involved with their projects' administrative and financial aspects, but not the detailed design. Some larger projects are also headed by a principal in charge, who serves as the in-house reviewer.

- Most civil design projects, even small- to medium-sized projects, involve several disciplines (e.g., geotechnical, structural, electrical, mechanical, and hydraulic engineering). Design within each discipline is headed by a lead engineer of that discipline.

Most states do not have explicit guidelines as to who should certify the drawings for all cases. The most explicit guidelines given by all of the states are that an engineer should not certify professional work outside of his or her area of expertise, and that the work that he or she certifies should be performed by him or her or under his or her direct supervision. These guidelines have several important implications:

- An engineer should only certify the portion of the design that is prepared by him or her or under his or her control and direct supervision. Therefore, a project manager who only manages the administrative aspect of the project, but not the technical aspect, should not certify the drawings. In addition, because the work is not performed under the direct supervision and control of a reviewer, he or she should not certify the drawings.

- If a set of construction drawings contains more than one engineering discipline and is prepared under the direct supervision of more than one engineer, then more than one engineer is allowed to certify that set of drawings.

Most states strictly prohibit the practice of *plan stamping*, which involves stamping the work of others with little or no review and with no supervisory role. This practice is illegal.

In some cases, the engineer responsible for a set of design drawings will leave an organization before certifying his or her drawings, and the organization will replace that former employee with an equally capable and qualified engineer, who will assume responsibility of the engineer of record for that project. In these cases, which are common in the consulting industry, state registration boards allow the new engineer of record to certify the work, provided he or she reviews all previous design work (including drawings, specifications, design calculations, design methods, and approaches) and, if necessary, changes the design based on his or her review.

Some states (e.g., North Dakota [State of North Dakota, 2000]) require engineers who perform pro bono design work to certify their professional work. The reason is that a professional engineer is responsible for his or her work regardless of the fees for that work. Therefore, the engineer performing pro bono work is as responsible and liable for his or her work as if he or she were being paid. However, some state professional registration boards have the Good Samaritan law, which indemnifies pro bono professional work for certain liabilities. Engineers who intend to perform pro bono work should consult with the Board of Registration for Professional Engineers for the state in which he or she intends to perform the work. He or she should also find out the limitations of any Good Samaritan laws within that state. For example, in North Dakota, the Good Samaritan law only applies to emergency situations such as natural disasters.

12.3   Electronic Stamp and Signature

Some engineers allow their professional engineering stamp and signature to be scanned electronically and added to electronic copies of their drawings. This practice is intended to be a time-saving procedure, especially when a set of drawings involves many sheets. When an electronic file of a drawing with the engineer's seal is submitted to a client, anyone can edit the drawing while the engineer's stamp and signature remain on the edited drawing. As a precaution, most states require that each construction drawing or its cover sheet contains the original wet-ink, dated signature. An electronic signature on the drawings is generally prohibited. In addition, when an electronic seal is used, some states require that additional statements be added to the drawings to prove that the engineer authorized the drawing. The state of Texas (1991), for example, requires the following statement when a *CADDSEAL* is used:

> *The seal appearing on this document was authorized by (name), P.E. No. _____, on _____ (date).*

The state of North Dakota (2000) provides the following guidelines for those who prefer to use an electronic seal:

1. The registrant must sign the original drawings.
2. Electronic copies of the drawings shall be submitted without the electronic stamp and signature.
3. The electronic copies of the drawings shall have a declaration stating "This document was originally issued and sealed by (name), (registration number) _____ on (date) and is stored at (location)."

The use of an electronic seal on drawings is hardly worth the effort. It does not save significant time for the engineer, and it can involve major legal issues and impact one's professional license. To satisfy some state board requirements, each drawing with an electronic seal must still be hand-signed and dated, along with an additional statement on the drawings regarding approval and authenticity.

CHAPTER 13

DESIGN CHANGES AND RECORD DRAWINGS

13.1   Design Changes

Rarely is a civil engineering construction project completed without some changes during construction. In some cases, it is necessary to change the design during bid solicitation, before construction has even started. One can look at a set of design drawings as a living document that continues to evolve during design, bidding, and construction. Regardless of the amount of changes required, each revision to the design drawings should be documented. Whether they are made during bidding or during construction, design changes are usually associated with changes in construction cost. This section discusses common reasons for design changes and their impacts on bookkeeping effort and cost implications.

Design changes made during bid solicitation are incorporated as amendments or addenda. Some of the reasons for making changes to the design during bidding are:

- Schedule considerations—When a construction schedule is tight, an owner sometimes advertises and distributes bid documents to initiate the bid process before the design is finalized, with the understanding that the finalized version can be completed by the time bids are due. This tactic may work in some cases, but is generally risky. Making significant design revisions during bidding is a practice that should be discouraged. Bidders, when presented with amendments with significant design revisions, should be given additional time to evaluate the revisions and adjust their bids accordingly. If no additional time is allowed in the amendments, the bidders will request additional time or will add substantial contingencies to their bids. Therefore, it appears the original goal of starting the construction on time may not be met using this tactic, depending on how complete the design package is at the beginning of bidding.

- Design omissions and errors—When design omissions and/or errors are discovered, whether by the design engineers or by the bidders during bidding, they can be corrected or clarified by amendments.

- Value engineering—It is quite common for bidders to suggest changes to the design, usually for their own benefit and advantage. Some owners do not allow value-engineering alternatives during bidding; the bidders must base their bids on the original design, and alternative value-engineering proposals will be considered by the owners after contract award. However, some owners allow value-engineering alternatives during bidding. The proposed alternatives can be submitted with the bid and evaluated by the engineer during bid evaluation. It is important that the contract documents indicate that the engineer's decision on a value-engineering alternative is final.

- Additional design and construction requirements—Because these additional requirements or data are not available during the design period, this reason is almost always related to the schedule. Additional requirements may include permits, temporary easements for construction, and laboratory test data.

Once the bid is awarded, it is best for the owner to issue a revised set of drawings and specifications that incorporate all changes made to the documents during bidding. This conformed set of construction documents should contain the following information:

- All addenda issued during bidding

- All revisions to the drawings and specifications

- A documentation of bidders' questions and the owner's responses to those questions

The conformed set then becomes the final documents for construction. When significant changes are made during bidding, a conformed set of documents for construction eliminates the confusion that may arise from interpreting a superseded set of drawings and specifications.

Design changes are almost inevitable during construction in most medium- to large-sized civil engineering projects because these projects involve subsurface conditions, underground structures, groundwater, and other uncertain site conditions (see Chapter 3). For example, the depth of the foundation may be adjusted for unexpectedly weak subgrade encountered during excavation, or the type of foundation may need to be changed (e.g., from shallow foundation to deep foundation) if a competent subgrade is not encountered during excavation.

Most design changes are associated with changes in cost, and sometimes schedule. When a design change is made during bidding, bidders adjust their bids accordingly. When a design change is made during construction, the changes in cost and performance period depend on the nature of the changes and provisions in the general conditions and supplemental conditions allowing for changes.

When a drawing is revised during design, bidding, or construction, it is the engineer's responsibility to properly document the changes. Any time a drawing is altered, a new version or revision should be designated for that drawing. The version number or revision number is shown on the title block of a drawing (see Section 5.2). Different engineers have different ways to designate revisions, and all of them are acceptable, as long as the version used is consistent for the entire project. The numbering (1, 2, 3, etc.) and lettering (A, B, C, etc.) systems are used commonly. It is important to point out that—because drawings not affected by design or field changes do not need to be revised—drawings from the same set can carry different revision numbers. When a drawing is revised, a short, abbreviated explanation should be used in the revision block of the title block to accompany the revision number:

Revision no.	Description
2	Bid Amendment No. 1
3	Change Alignment and Inverts of 18-inch Pipe

It is customary for designers to use a "cloud" symbol around design features on the drawings that are affected by the change. A triangle with the revision number is placed adjacent to the cloud symbol. This practice is acceptable when the added symbols do not overly complicate the graphics of the drawings. When significant revisions are made in a drawing, it is recommended that the cloud-and-triangle symbol system be omitted for the sake of simplicity. Instead, the revisions can be replaced with notes describing the changes on affected drawings. More elaborate descriptions of the design changes can be documented with a technical memorandum or a letter.

13.2   Record Drawings

Record drawings are drawings issued after construction of a project is complete. The term *as-built* (formerly used to describe these drawings) is less desirable from a professional liability standpoint because it implies that the drawings represent the exact configuration of all construction features. When an engineer does not have a full-time inspector on site during construction, it is impossible for the designer of record to certify that everything shown on the plans and specifications has been installed in accordance with approved plans and specifications. In fact, even when the engineer has full-time involvement during construction, it is still not possible to claim that everything shown on the plans and specifications has been installed. Because of this difficulty and uncertainty, the term *record drawings* is preferable because it implies that the drawings thus prepared are based on the engineer's observations and records on file during construction.

Record drawings should show the changes made in the design during bidding and construction. These records are typically maintained by the engineer. Some owners require the contractor to maintain a set of marked-up drawings (*red-line drawings*), showing all current changes made to the design to date, on site during construction. At the end of construction, the contractor is required to submit the red-line drawings as part of the close-out documents. The engineer then prepares the record drawings based on the red-line set and all of the construction records in his or her file. The complete record drawings set should be assigned a new revision number with the designation "Record Drawings" on the revision block.

PART 3

TECHNICAL SPECIFICATIONS

CHAPTER 14

PURPOSE AND USE

14.1 Role of Technical Specifications

Technical specifications are written instructions and requirements that accompany construction drawings. The term *specification* is used interchangeably with *technical specification*. Used together, the specifications and drawings comprise all of the technical construction requirements to complete a project. They are part of the contract documents. As such, they carry certain legal implications. As discussed later, specifications and drawings should not contain duplicate information; however, in the event of a conflict between drawings and specifications, the provisions of the specifications usually take precedence. Therefore, the preparation of specifications should receive as much attention as that of drawings.

In general, specifications contain all necessary information that is not shown on the drawings. Specifically, specifications for a civil design project contain the following information:

1. Detailed material requirements, including quality standards

2. Testing requirements for quality control and quality assurance

3. Procedures for installation or placement of materials and equipment, including tolerances

4. Schedules or lists of materials or equipment otherwise not shown on the drawings

5. Coordination of work among different trades and disciplines, including restrictions, conflicts, and limitations

6. Submittal and schedule requirements

7. Measurement and payment provisions for all work items

8. Miscellaneous general requirements (e.g., environmental abatement, safety) that cannot be depicted on the drawings

9. Permits obtained by the owner

10. Reference data such as climatic data, stream flow records, field and laboratory test data, and records of existing site and facilities

11. Coordination with other contractors on site

12. Safety issues and responsibilities

14.2   Users of Specifications

For the most part, specifications are written for the constructor (contractor) and his or her subcontractors, fabricators, and material and equipment suppliers. The other two parties of a construction contract (the owner and the engineer) also use the specifications. The following is a description of how these three parties use the specifications.

Contractor

A contractor uses specifications when he or she expresses an interest to bid on a project. Based on the scope of work required in the specifications, the general contractor assembles a team of subcontractors, material suppliers, and equipment manufacturers. The first user within the contractor's organization is the cost estimator, who is responsible for bid preparation. In preparing his or her bid, the cost estimator uses the material and equipment requirements to obtain quotes and determines what contingencies to put in the bid for uncertain and risky factors. He or she also determines what labor categories and construction equipment are suitable for the project and arrives at the labor and equipment costs.

It is, of course, during construction that the construction crew uses the specifications. In the contractor's office, the manager prepares the required submittals (e.g., materials test data or shop drawings), orders the materials and equipment, schedules the required crew and equipment, and handles progress payments. In the field, the superintendent directs his or her crews, and sometimes his or her quality control testing personnel (depends on contractual arrangements), and receives materials and equipment delivered to the site. He or she interacts with the resident engineer to confirm or inquire about specified requirements. Compliance with permit requirements in the specifications is also a task for the contractor.

Material suppliers and equipment manufacturers or distributors furnish products based on the specifications. Compliance with the specifications is usually handled through the submittal process and testing.

Owner

There are several reasons that an owner is interested in the contents of the specifications. The first reason is that an owner needs to know what he or she is buying for the money he or she is spending. Secondly, an owner may have certain preferences (e.g., for a brand name with good past performance or consistency with the spare parts inventory) regarding products or end results. Thirdly, for operation and maintenance of the new facility, an owner will need a record of what is being built.

Engineer

Typically, the owner hires the engineer who prepares the specifications to engineer and manage construction. This arrangement is frequently the case in private-sector construction. Some government agencies have their own construction management teams, and the design engineer is hired during construction to perform certain technical tasks only. Regardless of the arrangement, the resident engineer has the responsibility of overseeing enforcement of the specifications during construction. He or she interacts with the contractor superintendent and foremen, directs his or her field inspectors and quality assurance testing personnel, communicates with the designer, checks the materials and equipment delivered to the site, and verifies permit compliance. It is important to consider the resident engineer's authoritative limitations. He or she does not have the authority to direct the contractor. For unacceptable work, the resident engineer should inform the owner, who is then responsible for notifying the contractor to fix the problems and to comply with the specifications.

14.3   Relationship with General and Supplemental Conditions

The general conditions and supplemental conditions define the duties and responsibilities of the owner, the engineer, and the contractor regarding all contractual issues (administration, technical issues, bonds and insurance, payment, dispute resolution, deliverables, and submittals). Technical issues in general and supplemental conditions are handled differently from document to document and from owner to owner. The general conditions by Engineers Joint Contract Documents Committee (1990), for example, define the guidelines for items related to technical specifications as follows:

- Reference to standard specifications (such as American Society for Testing and Materials [ASTM], American National Standards Institute [ANSI]) and specifications of technical societies (such as American Concrete Institute [ACI], American Water Works Association [AWWA]). Version of these standards is defined, and guidelines are provided to resolve conflicts between these reference specifications and the contract specifications.

- Definitions of the contract parties referenced in the specifications. These parties include the owner, engineer, contractor, subcontractor, supplier, and manufacturer.

- Meanings of certain terms used in the specifications that are open to interpretation. These terms include *as directed, as allowed, to the satisfaction of, acceptable to,* etc.

- Definition of changed conditions.

- Contractor's responsibilities on safety and his or her labor, equipment, and materials.

- Handling of substitutes and items pertaining to *Or Equal* provisions.

- Conditions of the site and site cleanliness.

- Submittal procedure and handling of shop drawings.

- Owner/engineer's testing and inspection.

- Protocol on rejecting, correcting, and accepting defective work.

- Warranties.

Because the general and supplemental conditions contain so many references to specifications and drawings, it is important that the specification writer is thoroughly familiar with these contract documents so that the technical provisions are consistent with these documents. It is vital that a knowledgeable reviewer checks all cross-references between documents.

14.4   Relationship with Drawings

Specifications and drawings work hand-in-hand. The correlation between specifications and drawings has been compared to a dovetail joint in woodworking, a union of two parts that complement each other and fit together with no overlaps and no gaps (Construction

Specification Institute, 1996d). Some general correlations between specifications and drawings are listed below:

- Consistent terminology should be used in specifications and drawings. Inconsistent terminology between the documents results in confusion, errors in construction, disputes in acceptance of work and payment, changed conditions, and claims. Some examples of inconsistent terms used in drawings and specifications are:

  Material Types and Classifications
  In earthwork, when a material called *random fill* is used in the drawings, the same term should be used in the specifications. This term should not be called *compacted fill, fill, backfill* or other names in the specifications. This distinction is particularly important in a large earthwork project, when several fill materials are required. Each fill material will have a designation, such as *general fill, random fill, embankment fill, structural fill, select fill*, etc. In concrete work, when a material called *backfill concrete* is used in the drawings, this term should not be called *mass concrete* or *lean concrete* in the specifications. In mechanical work, a *sluice gate* shown on the drawings should not be called a *slide gate* in the specifications, because these two types of gates have some significant differences in quality and performance.

  Facility and Project Feature Components
  When a *ditch* is shown on the drawings, the specifications should not refer to this feature as a *trench*. When a *cofferdam* is called out on the drawings, this feature should not be referred to as *protection barrier* or *enclosure* in the specifications. When a *spillway* is shown on the drawings, that feature should not be called *sluiceway* or *outlet* in the specifications. When a *chamber* is used on the drawings, that feature should not be referred to as a *vault* or *adit* in the specifications.

- There should be no duplication of information on the drawings and specifications. Strictly speaking, there is no harm in duplicating exact, consistent information on both documents. However, this practice may result in problems when that information is changed during design, bidding, and/or construction. Unless identical changes are made in both documents, conflicts and discrepancies will be the result.

- Specifications are used to give a detailed description of materials and equipment, and that function should not be repeated on the drawings. It is a good practice to call out materials or equipment on the drawings by names or designations, but reserve the specifications for its full details. This practice avoids excessively long callouts or notes on the drawings (see Section 8.3) while allowing the written format of the construction documents (the specifications) to describe the required quality and performance. Some examples are given below.

  Subgrade Preparation
  When subgrade preparation is needed below a structure, the drawing should only show the limits, while other preparation requirements can be handled by calling out "subgrade preparation, see specifications." In the specifications, subgrade preparation requirements of depth of scarification, moisture conditioning, proof-testing, and compaction can be described in details.

  Construction Joint
  When a construction joint is needed in a cast-in-place concrete structure, the drawing should only show "C.J.," which is a commonly used abbreviation for construc-

tion joint. In the general notes of the drawings, a reference to the specifications can be used for all construction joints. In the specifications, construction joint requirements of cleaning, sand-blasting, high-pressure water jetting, and criteria of acceptance can be described in detail.

- Specifications can be used as an extension of the drawing features that require excessive amounts of text descriptions. Drawing features with excessive amounts of text include schedules of materials or equipment and notes. From a production standpoint, it is much faster to produce or change these schedules and long notes with word processing software than with CAD software.

The timing of specifications preparation during design is important in obtaining a quality design product. In the production of construction documents, it is common for designers to start with drawings, which require a significant amount of the design effort. On the contrary, specifications preparation is usually relegated near the end of the design process, as if the document is an afterthought. Some of the reasons that designers give for this practice are:

- Specifications cannot be completed until most of the drawing features are identified and drawn.

- Writing specifications is not pleasant, and given a choice, a designer would rather engage in a more pleasant activity, such as drawing preparation.

Specifications that are prepared in a hasty manner contain errors, omissions, discrepancies, and inaccuracies that impact bid prices, result in serious construction problems, and cause claims. Numerous court cases demonstrate that when a conflict arises between drawings and specifications, the court usually rules favorably toward the specifications (Fisk, 1992). According to Federal Acquisition Regulations (FAR) for contract construction, the preference for specifications over drawings is standard. So, regardless of how perfect the drawings are, a defective set of specifications accompanying these drawings will cause serious problems on a project.

Under an ideal situation, specifications should be prepared concurrent with drawing production. However, there are limitations within the design industry—man-power availability, schedule constraints, and budget constraints—that make it difficult to achieve this goal. Rather, the following compromise approach is suggested:

1. Using project and design requirements as a basis, establish a draft Bid Schedule at the start of a project.

2. Prepare a preliminary list of drawings.

3. Prepare a list of technical specification sections.

4. Start detailed specification preparation when the drawing production is 60% to 70% complete. At this stage, all of the design features, materials, equipment, and other project requirements are identified, thus allowing detailed specifications to be developed.

When referencing drawings in the specifications, the specification writer should resist referring to detailed locations of the drawings (e.g., "as shown on Section A of Sheet 7," or "as shown on Detail 6 of Sheet 10"). Drawing sheet numbers and section and detail designations and locations frequently are changed and moved during the course of drawing production. Therefore, a constant updating effort will be required in the specifications. The suggested practice in referring to drawings is to state simply "as shown on the drawings." It is the responsibility of the contractor to locate the feature on the drawings.

CHAPTER 15

TECHNICAL AND DESIGN ISSUES

15.1　The Specification Writer

This section addresses the question: Who should write the specifications? For this discussion, the person who prepares the specifications is called *specifier*. Traditionally, technical specifications are prepared by the following personnel categories:

1.　Design engineers who are also involved with engineering analysis of the project and production of the construction drawings. Depending on the different disciplines for a particular project, more than one engineer may be involved.

2.　In-house specification writers whose sole responsibility is to prepare specifications. These are professionals who have certain basic training and experience in writing specifications, but who are usually not involved directly in the analysis and production of construction drawings. They may or may not be engineers. This approach is usually used in large engineering firms or large government agencies such as the U.S. Army Corps of Engineers or the U.S. Bureau of Reclamation.

3.　Specification writers contracted from outside the engineering firm. The qualifications and experience of these outside specifiers can be assumed to be equal to, or better than, those of the in-house specifiers.

4.　Staff engineers who have no previous involvement with the project or have little experience in specifications writing, but are assigned this task because they are available and are lower-paid staff.

The best specifier for civil engineering design projects should fit into Category 1 (above). Design engineers understand the technical requirements and design intent for projects, are familiar with site conditions and site constraints, and coordinate well with construction drawings because they are involved directly with production. It is important that specifiers work within their areas of expertise. In fact, all state Boards of Registration for Professional Engineers (see discussion in Chapter 12) require this. For example, specifications on excavation, dewatering, and earthwork should be prepared or supervised by geotechnical engineers; and specifications on structural concrete and fabricated steel should be prepared by structural engineers. Another important qualification for an engineering specifier is field experience. A specifier with little or no field experience, no matter how technically qualified, is likely to produce specifications that may not be constructible or may be so difficult to construct that the cost will increase. Lack of field experience in construction specifiers is one of the most common complaints from contractors.

In large organizations, especially for large projects, technical specifications are prepared by in-house professional specifiers that fit into Category 2. This arrangement has been used successfully for many years. Some of these specifiers do not have engineering training, but most of them have vast construction experience, either as contractors or construction managers. Some of them are certified by professional organizations, such as Construction Specifications Institute. Assuming these in-house specifiers are qualified to prepare specifications, the success of producing a set of well-prepared project specifications should include the following work on the part of the specifiers:

- A visit to the project site to become familiar with site conditions and site constraints

- Meetings with lead design engineers to gain an understanding of design criteria, design intents, and testing requirements

- A review of partially completed or nearly completed construction drawings

- A review and understanding of the basis of the bid schedule

- Obtaining from the design engineers a list of materials, product data, or material sources

- Agreeing on terms and names to be used in the drawings and specifications

The main advantage of using in-house professional specifiers is that the design engineers can continue to focus on designs, design drawing production, and design report preparation. This in-house resource, when properly used, is a valuable asset in a design-oriented organization.

The considerations for outside professional specifiers that fit into Category 3 are similar to those of in-house specifiers, except in that they are brought in near completion of the project. When the schedule of producing the specifications becomes critical, there is a tendency on the part of the designers to limit some of the information available to the specifiers. Therefore, it is up to the specifiers to request additional information during the production process. When adequate time is available and the budget allows for it, engaging outside professional specifiers should achieve similar success as that of in-house specifiers.

Using specifiers that fit into Category 4 should be avoided. In the long run, the cost of producing a set of well-prepared specifications will likely increase because of significant changes required during the review process. In the worst scenario, errors in the specifications will not be detected until construction, resulting in cost overruns, claims, and disputes.

The American Society of Civil Engineers Committee on Specifications of the Construction Division conducted a questionnaire survey of more than 600 private owners, design professionals, and municipal, state, and federal agencies regarding various topics related to specifications (ASCE, 1979). One of the questions—what minimum qualifications should a specifications engineer possess?—is relevant in this discussion. The following is a list of qualifications provided by the respondents, in order of decreasing importance:

1. Field (construction experience), frequently expressed in the 5-year to 10-year range, with a few suggesting resident engineer experience

2. Engineering degree

3. Professional registration

4. Design experience

5. Design and field (construction) experience

6. Technical writing ability

7. Prior knowledge or understanding of project

8 Legal understanding, knowledge, or education

9. Knowledge of materials and their availability

10. Fluency in English and ability to communicate

11. Specification writing experience

12. Two years of college education

13. Common sense

14. Knowledge of Construction Specifications Institute

15. Formal training in specification preparation

16. Project manager

17. Master's degree

18. Familiarity with state standards or local codes

19. Maintenance experience

20. Research ability

21. High school graduate

22. Principal of firm

It is interesting that the "Principal of firm" is listed dead last, behind "High school graduate." From the perception of owners, the principal of a firm is least qualified to write technical specifications. Some of this perception is unfounded, as some principals are well qualified with sound field and design experience and technical backgrounds.

15.2   Problem Areas

Many claims and construction disputes arise from problem areas associated with technical specifications. This section first identifies what most of the problem areas are, and the remainder of this section and Chapter 16 discuss how to mitigate these problems. The following is a list of problem areas in specifications, in no specific order, that could lead to disputes, claims, and/or litigation:

Technical Inaccuracy

This problem area pertains to inaccurate technical data in the specifications and deficiencies in product performance requirements. The sources of this problem may not be self-evident without some investigation on the nature of the inaccuracy. When a design engineer makes an error in analysis or design, inappropriate material will end up in the specifications, or an inaccurate design feature will end up in the drawings. Deficiencies in product performance can result from design error or from a defective product from the manufacturer. Some of these problems are not manifested until after construction, when the new product or material is put into use, or when the facility has been operating for some time.

Examples of technical inaccuracy in the specifications include:

- A laboratory concrete mix design contains defective data, and the results cannot be reproduced by the contractor in the field

- A design engineer makes an error in selecting the thickness or type of a pipe, and the pipe ultimately deforms excessively under the design load

- A design engineer makes an error on the gradation of a sand and gravel filter, and the filter ultimately fails and leads to inadequate seepage collection and filtering

- A geotechnical engineer does not adequately characterize the foundation of a site, resulting in extra cost in foundation dewatering and temporary foundation support during excavation

- A Type I cement is specified for a cast-in-place concrete structure below grade, even though laboratory testing indicates that the foundation soil is prone to severe sulfate attacks, and a Type II or Type V cement is required

- An aluminum handrail is bolted directly to a steel-frame decking, resulting in serious cathodic corrosion problems

- A new product or new construction technology is specified, and the performance does not meet the advertised guarantee

Product Substitution

One of the most litigated problems in construction is related to product substitution provisions. Specifically, the controversy is on the interpretation of one of the many versions in product specifications such as "or equal," "or approved equal," and "or approved equivalent." In federal construction projects, proprietary specifications (see Section 17.7) cannot be used unless they are followed by "or approved equal," or some such allowance to encourage fair competition. When a product is specified with one or more brand names plus a provision to allow the contractor to substitute an equal product, the contractor is given a choice. A problem arises when the contractor's choice and the engineer's preference are different. The controversy continues to be a problem, in spite of the fact that all contract documents contain explicit policy regarding product substitution.

Regardless of which side is right or wrong, disputes on product substitution that involve "or equal" provisions are based on the following circumstances:

- During bidding, a bid is based on a substituted product that is later rejected by the engineer

- During construction, a specified product with a particular name is no longer available, and the contractor and engineer cannot agree on an equivalent product for substitution

Mitigation of this problem area is discussed in Section 15.6.

Ambiguities and Conflicts

During construction, ambiguities and conflicts in the specifications cause disputes between the contractor and the engineer. An ambiguous specification requirement has more than one meaning and interpretation. Examples of ambiguous language include:

- *The minimum thickness of the fill shall be 12 inches.* The contractor placed 36 inches of fill, which met the specifications, but the owner only intended to pay the contractor for 12 inches of fill.

- *The Contractor shall submit rebar shop drawings to the Engineer for review and acceptance at least 10 days prior to forming the structure.* The contractor interpreted that 10-day period as 10 calendar days, when the engineer's intent is 10 working days (which is 2 weeks), resulting in a difference of 4 days in the submittal schedule.

- *The Contractor shall dispose of all waste materials off site.* The engineer considers excavated earth materials to be included as waste materials, but the contractor does not, and the contractor demands additional compensation to haul the excavated earth materials off site.

In litigation on ambiguous specifications, courts have interpreted strongly against the party that prepares the specifications—the engineer. Therefore, the burden of avoiding ambiguous language and requirements is on the engineer, who is the owner's agent. Conflicts arise when there is a discrepancy between the drawings and specifications on the same requirement, or a discrepancy in the specifications for the same requirement. When there is a conflict, several outcomes are possible:

- A contractor requests a clarification from the engineer, and the engineer provides the contractor with the proper clarification.

- The contractor deliberately makes an interpretation on his or her own that is in his or her best interest (i.e., at the lowest cost to him or her).

- Both the engineer and contractor are unaware of the conflict, and an unintended error is made in the construction.

- Not all conflicts lead to disputes. Examples of conflicts in the specifications that may lead to disputes and claims include:
  (a) A pipe flange is called out with a rating of 150 pounds on the drawings. In the specifications, a standard American Water Works Association (AWWA) reference specification is used, and a different class of flange is specified.
  (b) A 10-foot-high 0.5H:1V temporary excavated slope in sandy soil is shown on the drawings, and the contractor is also required to comply with Occupational Safety and Health Administration (OSHA) safety requirements. According to OSHA excavation regulations, the slope shown on the drawings is too steep for that type of material.
  (c) A concrete mix is specified using the descriptive approach (see Section 17.2; i.e., all proportions are specified) while limiting the slump to a consistency that can be pumped. The contractor furnishes the concrete as specified, but the consistency is too stiff for pumping.

Recommendations to mitigate ambiguities and conflicts are discussed in Section 16.2.

Constructibility

The constructibility problem is related to whether a specified end result can be achieved using customary industry means and methods. Unconstructible design is a direct result of inexperienced designers and unqualified specifiers who lack adequate construction field experience. It is also caused by inadequate investigation of a particular site during design, which limits the types of construction possible for that site. Examples of unconstructible design include:

- Tolerances that are too strict to be practical. It is not reasonable to specify tolerance for grading or for laying a drainpipe to within 0.01 foot. Earthwork tolerance is generally limited to within 0.1 foot.

- A no-tolerance approach to field quality control is used, particularly on earthwork. It is difficult to expect a contractor to adhere to 100% compliance of earthwork (e.g., regarding specified gradation, compacted density, and in-place moisture content). This unreasonable tolerance is one of the major contractor complaints.

- A compacted fill, placed in 12-inch lifts, is required directly above a soft clayey foundation. Even with low-pressure equipment, the contractor cannot be expected to compact the fill, because of the soft subgrade.

- A tunnel-boring machine (TBM) is specified for constructing a tunnel in a site known to contain shear zones, clay gouge, and breccia zones. Because of concerns that the TBM will be frequently stuck in the bore with frequent down time and loss production, the contractor chooses another method of tunnel excavation. In the worst case, the TBM can be permanently lost in the bore and cannot be retrieved.

Unenforceable Requirements

Whether it is intentional or not, some specifications contain provisions and requirements that cannot be enforced during construction. Most of these provisions are called *open targets* that are unbiddable. Examples of unenforceable requirements include:

- *As directed by the Engineer, to the satisfaction of the Engineer, as determined by the Engineer.* Without acceptance criteria or provisions to compensate the contractor (e.g., establishing a unit price for amount above bid schedule quantity), the amount of work required to please the engineer is unknown, and therefore subject to dispute.

- *The Contractor shall provide cofferdams, pump, etc., for proper diversion of stream flows away from the construction site.* There are two open targets here: the words *etc.* and *proper.* The contractor understands that he or she needs to provide cofferdams and pumps to divert stream flow, but does not know what other requirements are necessary. Therefore, this provision is unbiddable. The term *proper* is subjective. What is considered proper to the contractor may not be considered proper to the engineer or to the owner. In fact, other subjective terms such as *good, workmanlike, adequate,* and *enough* should also be avoided for similar reasons.

Avoidance of unenforceable requirements through good specification writing is discussed in Sections 15.9 and 16.2.

Inadequate Attention to Specification Production

The production of technical specifications for a project typically occurs near the end of the project, and not enough attention is given to its production (see Section 14.4). Whatever the reason for this common practice—whether a resource issue, project budget issue, or sequence of construction document production—the lack of proper attention in preparing a set of well-written specifications may result in a suite of problems. Some designers are anxious and enthusiastic in preparing construction drawings, but do not look forward to working on specifications. It is ironic that the document that receives inadequate attention during design receives precedence over the drawings in the event of a conflict. It is well known that courts generally rule favorably toward typed documents (e.g., specifications) over printed documents (e.g., drawings). With so much at stake for successful completion of a construction project, the design profession should at least give equal priority to specification preparation as to construction drawings and elaborate upfront analytical work.

15.3   Philosophical Approach

Over the years, the author has developed an approach that has worked well for all parties involved in developing technical specifications for civil engineering construction projects. His approach bears some resemblance to the *partnering concept* that has been promoted recently, especially in public-sector work. It affects the disclosure of information in the documents, the sharing of risks among the parties, and fair treatment of the contractor, all while protecting the best interests of the engineer's client, the owner. The following is a discussion of the main philosophy for this approach:

The Most Competent Contractor does not Always Get the Job

This attitude is used for projects that are bid on a competitive basis, especially for public-sector projects, in which the government is sometimes required to award the contract to the lowest bidder. This is not to say that the lowest bidder will necessarily be an incompetent contractor. When one assumes that a problematic contractor may be involved with the construction, there are certain loss-prevention measures that the designer should use during design to minimize claims, disputes, and delays:

- There can be no obvious flaws in the specifications that the contractor can use as justification for claims. This requires considerable experience in specification writing. There should be no technical inaccuracies (see Section 15.2). No ambiguous language should be used that will allow the contractor to take liberties in his or her own interpretation. There should be no conflicts within the documents (specifications and drawings) that the contractor can use to his or her advantage.

- Clearly spell out the duties and responsibilities of the contractor, such as submittal contents and schedule, quality control testing and frequency of testing, utility clearance, progress meetings, all safety issues, and progress payment requests, etc.

- State in corresponding sections all minimum qualifications and experience for certain special construction, such as anchors, roller compacted concrete, and foundation improvement. This protects the owner from getting inexperienced contractors that get the job through low bid.

The precautions listed above are nothing more than what the owner normally expects from the design engineer in preparing a good set of specifications. By maintaining this defensive approach throughout the preparation of design documents, the designer can guard against potential problems that could manifest themselves during construction. This approach is no different from the standpoint of the owner in engaging a professional engineer, who is required to carry certain professional liability insurance in the event of negligence, and to perform his or her work to comply with an industry standard of care and a scope of work prepared by the owner.

The author believes that there is great integrity, professionalism, and expertise among most construction contractors, and he has the greatest respect for contractors who frequently have to deal with difficult working conditions and risks to complete the work. He has learned a lot from competent contractors in his career, and continues to value contractors as a great resource for design and construction issues.

Sharing the Risk

One of the civil engineer's first responsibilities, besides public health and safety, is to protect the interest of his or her client, usually the owner. In heavy civil construction, there are many times when the engineer and the contractor have to deal with unknown conditions, such as inclement weather, differing foundation conditions, groundwater inflows, flooding, and stream diversion. One apparent way to protect the owner in these situations is to shift all the risk to the contractor, as in the following specifications:

> *The Contractor shall provide whatever is deemed necessary for cofferdam protection, stream diversion, and dewatering so that all construction along the stream channel is performed in the dry. Damages to permanent construction caused by flooding shall be repaired at no additional cost to the Owner. It shall be the responsibility of the Contractor to determine the height of the cofferdam and size the diversion facilities for his or her own protection.*

In this particular example, the unknown factor is the level of flooding during construction. Even when the engineer provides the bidders with historical stream flow records and seasonal precipitation records, there is still considerable risk for the contractor to work in the stream channel. Bidding on unknown conditions and risky situations is difficult, and, as part of the bidding strategy, potential claims may be prepared already by bidders. In general, unknown factors increase construction costs. If the increased costs are not manifested in the initial bid, they will most likely be manifested in change orders and claims during construction. Contractors are known to take risks, and some accept more risk than others. There is undoubtedly a cost difference in favor of the riskier contractor, but the owner is ultimately at risk if a flood destroys the work.

It is unfair for the contractor to assume all of the risk for unknown or unanticipated conditions. The owner should share appropriate risk with the contractor, who the owner hires to help build his or her project. When an owner agrees to share risk, the contractor feels that he or she is being treated fairly and that he or she is included as a partner in the team. It promotes a mutually acceptable working relationship between the owner and the contractor, and between the contractor and the engineer during construction.

Sharing the risk and treating the contractor fairly requires compensating the contractor for the costs and losses he or she experiences as a result of unpredictable conditions. To handle anticipated unknown situations during bidding, the bid documents can ask the bidders to put aside allowances, or some other uniform basis can be specified for bidding. In addition, the specifications should also allow the contractor to recuperate losses for documented unanticipated conditions through a negotiation process. When bidders enter their bids with the understanding that they will not need to absorb unnecessary losses, there will be less risk costs, and the bids will be more uniform and consistent. When a contractor starts construction with the understanding that he or she will be protected from losses caused by unknown situations, there will be less adversarial confrontation and more cooperation during construction.

For the example given above, the specifications can be rewritten to provide a more uniform basis for bidding and an avenue to compensate the contractor:

> *The Contractor shall provide temporary cofferdam protection and stream diversion for construction in the stream channel. The required level of protection is elevation 362.0 feet. Temporary protection shall be earthfill cofferdams, sheetpiles, bulkheads, or other approved water-retention barriers to prevent flooding of the new structure during construction.*
>
> *Damages to the permanent construction and temporary protection facilities caused by inadequate protection below elevation 362.0 feet shall be the responsibility of the Contractor. The Owner will negotiate with the Contractor in accordance with provisions in the contract documents to compensate for damages for floods that overtop the cofferdam above elevation 362.0 feet.*

Fairness to all Parties

A fair specification treats all parties on the same basis. Each party has a certain responsible role, whether financial, design, testing, or construction. As designers, the engineer should take a lead technical role to provide the owner with the most cost-effective product. To the maximum extent possible, all of the design decisions and technical decisions should be made during the design phase, and not delayed until construction. One of the most common complaints from contractors is that they are asked to do too much engineering. With some exceptions of specialty contractors, most contractors think their main contribution to construction is to build, and not to design, for the owner. At the same time, owners pay engineers for their technical and design expertise, and a decision to delay a design issue to construction or to shift the design to the contractor should not be made, except under special circumstances. Special circumstances may include unknown field conditions that can only be revealed during construction, or improved cost-effectiveness by hiring a specialty contractor to design-build a certain product.

The following are examples of unfair and fair specification requirements:

- Fair to all—*The Contractor shall submit test samples from proposed source for testing by Owner. If the test fails, the Contractor may submit a recheck sample for additional testing by the Owner. If the recheck sample fails, then the stockpile or proposed source shall be rejected. Testing for recheck sample shall be at the expense of the Contractor.*

- Unfair to contractor—*The Contractor shall be responsible for providing adequate quantities of structural fill from on-site borrow areas furnished by the owner.*

- Fair to contractor—*The Contractor agrees that, should he or any of his employees in the performance of this contract discover evidence of possible scientific, prehistoric, historic, or archaeological data, he shall notify the Contracting Officer immediately. Where appropriate by reason of a discovery, the Contracting Officer may order delays in time of performance, or changes in the Work, or both. If such delays, or changes, or both, are ordered, the time of performance and contract price will be adjusted in accordance with the applicable clauses in the Contract.*

- Unfair to contractor—*The Contractor shall perform all concrete and field density testing at a frequency as deemed necessary by the Engineer.*

- Fair to all—*The Contractor shall perform concrete testing at a frequency of every 30 cubic yards of concrete placed. The Owner will perform all field density tests at least once every lift of earthfill placed, or more frequently, as deemed necessary by the Engineer.*

15.4 Technical Correctness and Quality Control

The first responsibility of a specifier is technical correctness of the specifications. A set of specifications can be written in the most clear and concise language, but if the wrong materials or products are specified, or if the wrong standard references are cited, the specifications are defective. Examples of technical inaccuracies are illustrated in Section 15.2. Technical inaccuracies are considered design errors. Problems associated with design errors may or may not be readily revealed during construction.

Ensuring technical correctness is one of the main functions of the lead design engineer and represents part of the engineering process in a design. Regardless of what engineering discipline (structural, mechanical, geotechnical, etc.) is involved, and without going into technical details in a particular discipline, there are a number of things a lead design engineer can do regarding technical correctness in specifications:

1. The design analysis, whether hand-calculated or computer-assisted, should be checked for correct assumptions, criteria, methodology, and mathematics. Unchecked analysis is a poor engineering practice, and the direct implications of an error in specifications caused by unchecked analysis can be serious.

2. When an old specification is used for a current project, the engineer should try to understand what design criteria, site conditions, and other project-related background are associated with the old specification before it is marked up for the next project. Requirements in specifications were put in the previous project for a reason, and that reason may be different for the other projects.

3. Only factual field and laboratory data (such as drill logs, laboratory soil test data, concrete mix design study, and borrow-area test pit logs) should be included in the construction documents. The engineer's interpretation and evaluation, used in reports to his or her client, should not be used in construction documents. The contractor should provide his or her own conclusions and interpretations of these data. A somewhat different approach of furnishing the engineer's reports and interpretations to the contractor is discussed in Section 20.1.

4. Heavy civil construction often involves many different engineering disciplines, and a designer of a particular discipline should resist practicing beyond his or her area of expertise without consulting with engineers from other disciplines (see Chapter 12). For example, a structural engineer should not prepare a foundation preparation specification of a clay shale foundation that has slaking deterioration potential if not adequately protected. A geotechnical engineer should not prepare a specification on mechanical gates and valves, especially for special applications, such as pinch valves, free-discharge valves, and radial gates.

5. Specifying new products, new materials, and new construction methods that do not have adequate performance history or adequate testing should be resisted. On one

hand, an engineer should always be innovative and look for ways to reduce construction costs and to advance the state of the practice. On the other hand, a design engineer has a responsibility to minimize liability exposure for his or her company and himself or herself, and it is natural to take a fundamentally cautious approach in design. These two opposing considerations should be balanced on a case-by-case basis.

6. The designer should make it a habit to review the standard reference specifications referred to in the project specifications, especially when using a new or revised version in which subtle differences may exist (see Section 17.4). When products are specified from standard reference specifications, the availability of those products should be verified with manufacturers and distributors.

The first responsibility of an in-house design reviewer is to check on the technical correctness of the specifications. If the primary role of the project manager of a design project is administrative because of a lack of design experience, that manager should not be the technical reviewer of the specifications (or the construction drawings). The same can be said about principals without design and construction experience. The review by a qualified technical reviewer is the last line of defense for quality control before the project goes into bidding and construction. A list of a reviewer's responsibility for specifications is provided by Construction Specifications Institute (1996d).

Typographical errors are another overlooked aspect of quality control of specifications. Typographical errors, if undetected, may result in serious problems in construction. An extra word (*shall not* vs. *shall*) or different numbers may have significant cost implications associated with the error. Typographical errors, even though committed by the clerical staff, are the responsibility of the lead design engineer in the same way technical errors committed by junior engineers are the lead engineer's responsibility. Regardless of time constraints caused by a tight specification production schedule, it is the responsibility of the lead specifier to back-check all changes typed by the clerical staff and all cross-references to other specifications and drawings. This problem is particularly prevalent when an old specification from a previous project is used as a starting point, and the text is so voluminous that there is a tendency not to carefully read or understand all of the text.

15.5 Contractor's Means and Methods

The conventional approach in construction, whether civil or architectural, is to allow maximum flexibility to the contractor's means and methods that are not specified, as long as the final product or final results comply with the specifications. The reasons for this approach include:

- It allows the contractor to adapt his or her own equipment, crew, and procedure for the project.

- It encourages the contractor to be innovative and use his or her own experience and background.

- It alleviates the designer's liability on the construction method and allows the field quality control personnel to focus on the end product instead of the construction method.

A well-known application of this principle is site-safety provisions, which are the sole responsibility of the contractor. Site safety is discussed in more detail in Section 15.12.

It is within the realm of the designer, however, to put restrictions on acceptable means and methods because of specific project requirements. These restrictions should not be considered a problem if they are known during bidding and construction. The following examples illustrate how limiting the contractor's means and methods actually results in a better product or performance, or improved safety, in some situations:

Soil compaction
Only vibratory rollers are allowed to compact the clean filter sand and gravel.

Only sheepsfoot rollers are allowed to compact the cohesive embankment fill in the dam. Smooth drum rollers are not allowed.

Drilling method
To avoid hydraulic fracturing, only drilling with hollow-stem augers and without drilling fluids is allowed in the embankment dam.

To avoid damaging the dam, only diamond-bit coring in the concrete dam shall be allowed.

Dewatering method
Well-point dewatering shall be required for excavation in the clean sand and gravel foundation. Foundation dewatering by sump-pumping alone shall not be considered adequate.

Construction sequence restriction
The Contractor shall first construct the new access road to allow local residents continuous access across the construction site.

To allow uninterrupted stream flow downstream, the Contractor shall install the new culverts.

In the examples above, the geotechnical engineer has determined—based on the anticipated subsurface conditions, his or her understanding of material behavior, and his or her judgment—that certain construction methods are not acceptable. This engineering decision can be made during design or during construction. To the extent possible, the decision should be made during design to minimize changes in conditions and disputes during construction. By specifying these restrictions in the bid documents, the bidders can accommodate these restrictions in their bids. By delaying this decision until construction, the contractor may argue that these restrictions will result in additional costs to him or her. Both approaches may end up increasing the cost of the project (in the original bid or in change orders), and handling the changes during construction could lead to other problems, including delays or impacts to other related work, which then leads to additional cost increase.

15.6   Specifying Materials/Products

Even if the specifications are prepared by nontechnical personnel, the selection of a specified material or product should be the responsibility of the design engineer (see Section 15.1). Many factors—functional characteristics, practical concerns, material and installation costs, code requirements, compatibility with existing facilities, past performance, maintenance requirement and cost, and availability—contribute to the selection of a particular material or product. The Construction Specifications Institute has an excellent reference on product evaluation (CSI 1996e).

The problems associated with product substitution are discussed in Section 15.2. In particular, the specification provision *or equal* is the center of most disputes regarding product substitution. Removing the *or equal* provision, as proposed by Rosen (1999), is not the solution to this problem. In fact, for federal projects, this provision is mandated if proprietary products are used. Rather, if used appropriately, the provision *or equal* is an acceptable provision to allow a contractor to substitute a named product for another one.

Let us approach the problem from the standpoint of a contractor preparing his or her bid on a certain named product with an allowance for substitution. Other than the brand name, no other properties and acceptance criteria are given in the specifications. To gain an advantage, the bidder does not use the specified brand name, but substitutes a less-costly product. This bidder later is selected as the contractor, and when the contractor submits the cheaper product to the engineer for approval, the product is rejected on the grounds that it is an inferior product and is not equal to the specified brand name. When a project gets to this point, claims and disputes are the obvious outcome, and lawyers are usually involved.

Consider another scenario. This time, the specified brand name is accompanied by a list of relevant properties and performance criteria that are used to determine product equivalence. Under these conditions, the avenues of controversy are removed. Let us also assume that the contractor, in preparing his or her bid, has identified a cheaper product that satisfies all of the specified properties and criteria, and has used that cheaper product in preparing his or her bids. During construction, he or she submits a request for substitution, and the engineer accepts the substitution because it meets all required properties and criteria.

The latter approach contains two strategies in the use of the *or equal* provision:

- It encourages competition, while it gives the engineer a contractual basis to compare equivalent products.

- When a set of characteristic properties and criteria are specified with a named product, the additional information allows the bidders to look for equivalent, less expensive products for use in his or her bid. Without this information, the bidder is risking that his or her substitution will be rejected later.

Two examples of appropriate use of the *or equal* provision are provided below:

1. *Geotextile erosion protection mat shall consist of polymer nettings, with a fused, three-dimensional mat of sufficient thickness and void space to allow for soil filling. The material shall be ultra-violet stabilized with a minimum of 2 percent carbon black to resist sun-light degradation when exposed. The mat shall be resistant to biological and chemical degradation. Geotextile erosion protection mat shall be Tensar Erosion Mat TM3000, or approved equal.*

2. *Steel sheetpiles shall conform to ASTM A328-93, and shall have a minimum yield strength of 39,000 pounds per square inch. The minimum section modulus for each single section shall be 90 cubic inches, and the weight per square foot of pile shall be 35 pounds. The sheetpiles shall be PZ35 as manufactured by Bethlehem Steel, or approved equal.*

In the examples given above, the requirement of uniform bid basis is used to encourage competition. The characteristic properties given in the specifications allow the engineer to approve or reject any substitution without having to deal with any potential for claims.

Finally, the engineer should check on the availability of a brand name product that will be specified. Failure to do so during design may be justification enough for the contractor to request a price increase during construction for an alternate product to substitute for the unavailable product.

15.7   Contractor's and Manufacturer's Roles

Occasionally, a design engineer is faced with the need to use specialty construction (such as high-capacity post-tensioned anchor installation, or a biopolymer slurry trench) or a new or specialty product (such as Howell-Bunger valve or underwater epoxy coating) to meet project requirements. The design engineer can seek design and test data and performance history from his or her peers or from published literature. In some cases, the engineer needs to consult with specialty contractors or product manufacturers to obtain additional design information and specification requirements. In civil engineering design, this type of interaction is quite common, and there are many advantages to doing so:

- Specialty contractors can provide first-hand experience that is valuable in understanding the application, limitation, constructability, and cost.

- Specialty contractors and product manufacturers can provide valuable guidelines on practical considerations that should be incorporated into technical specifications.

Some advice is offered here to take advantage of this valuable design resource while keeping "an arm's length" from contractors and manufacturers to prevent any appearance of impropriety and claims of collusion:

- The suggestions and guidelines provided by contractors and manufacturers, including guide specifications, should be carefully reviewed and screened so they do not favor the contractors or manufacturers consulted. Restricting the work to a particular contractor or a particular manufacturer not only discourages open competition, it is not allowed in federal construction. The engineer should disclose this intention up-front, prior to consulting with these sources.

- The engineer should not put the contractors and manufacturers in a position that may exclude them from bidding on the project being designed. Improper involvement that may jeopardize the contractors and manufacturers includes disclosing too much project or "inside" information or referencing their names in construction documents.

15.8   Specifying Tolerances

Tolerances are allowable deviations from a specified value, such as thickness and grades. Tolerances also apply to the degree of compliance for a particular specified requirement, such as quality control testing. These two tolerances have widely different meanings, and are therefore treated separately.

Dimensional Tolerances

Dimensional tolerances are used to give the contractor a range of acceptable finished dimensions, such as thicknesses, elevations, or verticality. Different practical tolerances are associated with different construction methods and different materials. For example, the elevation of the top of a concrete wall can be built to within 0.01 foot (3 mm) of accuracy using conventional forming and finishing methods, and the elevation of a foundation subgrade or fill can only be built to within 0.1 foot (3 cm) using conventional earth-moving equipment. That is not to say that tighter tolerances cannot be obtained. Tighter-than-normal tolerances require special construction methods that increase costs. To ensure that the owner does not pay more than is necessary, it is important that tolerances are not specified more than is necessary to accomplish the project design objectives.

When a specified tolerance is so tight that it cannot be met, it is considered unconstructible. An example of an unconstructible tolerance is the requirement of percussion drilling of a vertical hole to within 0.25% deviation from the vertical; a one-degree tolerance would be more appropriate. Unconstructible tolerance does not prevent a bidder from bidding on work, and the contractor will likely request a relaxation of the tolerance during construction. In such a case, tolerance relaxation should be granted by the engineer, unless there is a specific reason that tight tolerance is needed. In the example above, if tight tolerance is needed to protect an underground facility from being damaged by drilling, then a relaxation of the tolerance may put the existing facility into jeopardy. At this point, a different engineering solution or construction method may be needed, which usually implies a change order and delay. Another example of a tight tolerance is to finish a riprap layer to within 0.1 foot (3 cm), similar to a typical earthwork finish tolerance. The finish tolerance for riprap rocks should be in the range of 3 inches to 6 inches (7.5 to 15 cm), depending on the maximum sizes of the rocks.

Construction experience is helpful in obtaining constructible tolerances. In special cases, consultation with contractors will be required during design.

Degree of Compliance

A 100% compliance of a specified requirement, such as the minimum compacted density and moisture for an earthfill, or minimum concrete strength, is difficult to achieve, and the enforcement of such a requirement is frequently a source of disputes and claims during construction. Recently, a statistical quality control approach, which accepts less than 100% compliance, is becoming increasingly common in heavy civil construction projects. This approach uses a *running average* of all or part of the measured test parameters instead of the actual measured test parameter at any given time. The statistical quality control thus allows some marginal cases to be accepted. To prevent gross noncompliance, it is reasonable to specify an absolute minimum acceptance criterion. When a certain test parameter falls below that level, the construction will be rejected, regardless of the running average. The running average method can also be used in conjunction with measurement and payment. When the running average falls below a specified number, then the contract price will be reduced, depending on the degree of noncompliance. This statistical quality control has proved successful in many heavy civil construction projects in the area of earthwork, concrete, and pavements.

The following is an example of using a running average for quality control testing of concrete from the Colorado *Standard Specifications for Road and Bridge Construction* (CDOT, 1999):

> *The concrete will be considered acceptable when the running average of three consecutive strength tests is equal to or greater than the specified strength, and no single test falls below the*

specified strength by more than 500 psi. A test is defined as the average strength of three test cylinders cast in plastic molds from a single sample of concrete and cured under standard laboratory conditions prior to testing.

   When the average of three consecutive strength tests is below the specified strength, the individual low tests will be used to determine the pay factor in accordance with Table 601-2. The pay factor will be applied to the quantity of concrete represented by the individual low test.

15.9   Engineer's Discretion and Control

In the old days, construction was completed with minimal drawings and specifications, and the engineer played a vital role during construction, including significant day-to-day engineering decisions being made and exercising significant discretion and control over the contractor's work. That was before litigation became a preferred way to settle claims and changes made during construction. Now, construction documents become voluminous, even for small projects. The contractor determines his or her means and methods, and the engineer is discouraged from "directing the contractor" during construction. Specification provisions such as *as directed by the Engineer, to the satisfaction of the Engineer,* or *as determined by the Engineer* are treated as taboo because of fear that these provisions imply that the engineer is overstepping his or her bounds, or that the engineer is imposing an open target that will cause the contractor to lose profit or accept unwanted responsibility for site safety. Some specification writers even suggest that these terms should be totally removed from specifications, thus taking away the ability of the engineer to make discretionary decisions during construction.

   The main intent of these specification provisions is to give the engineer some discretion and control during construction when it is not possible to determine what decision is appropriate during design. The contractor should have no problem working under the engineer's direction or doing work until the engineer is satisfied, as long as the contractor is compensated for his or her efforts. Most disputes occur when the contractor discovers that there are no provisions in the contract for him or her to be paid for additional work directed or determined necessary by the engineer. With that in mind, the following compromises are offered in lieu of a complete ban of the following phrases:

As directed by the Engineer—The specified work that is directed by the engineer should not apply to contractor's means and method, and it should not give the impression that the engineer is creating new work for the contractor outside of the contract scope. Some examples of situations in which *as directed by the Engineer* can be used effectively are:

- *The Contractor shall clear and grub all trees within the limits of disturbance shown on the Drawings, except for several large trees to be protected. The Contractor shall mark the trees to be protected, as directed by the Engineer.*

- *The Contractor shall overexcavate soft foundation subgrade and backfill with compacted structural fill, as directed by the Engineer. Overexcavation and backfilling shall be paid in accordance with applicable provisions for Unclassified Excavation and Placing Structural Fill.*

As determined by the Engineer—The specified requirement that is determined by the engineer should stay within the context of the contract scope of work, and the outcome for the engineer's field decision should not affect the contractor's work or compensation. Some examples of situations in which *as determined by the Engineer* can be used effectively are:

- *The Contractor shall be responsible for testing the in-place density and moisture contents of the embankment fill. Two tests shall be performed for each compacted lift. The locations of the tests for each lift shall be as determined by the Engineer.*

- *The Contractor shall excavate 10 additional test pits in the borrow area shown on the Drawings. The test pits shall be at least 20 feet long by 10 feet wide and 15 feet deep, and shall be logged by the Engineer. The locations of the test pits shall be determined by the Engineer during construction.*

- *The Contractor shall drill and grout the voids outside the 48-inch concrete pipe. The primary drill hole pattern is as shown on the Drawings. The need for secondary and tertiary drill holes shall be determined by the Engineer based on actual grout takes in the primary holes. Payment for drilling and grouting secondary and tertiary holes shall be in accordance with the unit prices bid in the schedule.*

To the satisfaction of the Engineer—This provision should be used sparingly and only for unscheduled repair work when the contractor is clearly in the wrong, such as when he or she accidentally damages an existing structure to be protected. Because scheduled work should be based on specific acceptance criteria or other specific standards, this provision generally should not be used for scheduled work. Because the extent and scope of unscheduled work is not known during design, the acceptance criteria cannot be determined. Examples of *to the satisfaction of the Engineer* being used effectively are:

- *The Contractor shall protect existing benchmarks during site excavation. Damages to the existing benchmarks by the Contractor shall be repaired by the Contractor to the satisfaction of the Engineer at no additional cost to the Owner.*

- *Protect excavated subgrade. Subgrade damaged by the Contractor shall be repaired by the Contractor as directed by the Engineer and to the satisfaction of the Engineer. Repair may include excavation of disturbed soil, and replacing the excavation with compacted fill. The Contractor shall not be entitled to compensation for repair work from damaged subgrade.*

15.10   Handling Unknowns and Changed Conditions

Most civil engineering projects have some type of unknown conditions that will not be apparent until construction. Unknown conditions can be grouped into the following categories:

1. Conditions that would affect the design concept. For example, the foundation may contain so many boulders that it would be impracticable to drive the specified pile foundation, and a different foundation would be more appropriate. This category can be classified as *surprises,* caused by inadequate field investigation.

2. Conditions that do not affect the design concept, but would have an impact on measurement and payment of scheduled work. This category can be labeled as *plan quantity variation.* Examples of plan quantity variation include foundation excavation, grouting of foundations, and demolition of buried structures.

3. Conditions that could affect the construction working environment. This category is called *acts of God* and includes conditions that are so out of the ordinary that the changes cannot be considered during design. Examples of acts of God include unusually cold weather in winter construction, large flood runoff for construction near a stream, and excessive precipitation.

4. War, terrorist sabotage, etc.

Unknown conditions can be handled during design in the following manners:

Unanticipated Conditions

By definition of this category, the impact of an unanticipated condition on a particular design is not considered or overlooked in the design. Design changes caused by unanticipated conditions should be handled through contractual procedures on changed orders that will most likely result in redesign, schedule delay, and an increase in project cost. It should be recognized, on one hand, that—because of unknown buried utilities and structures, complex geology, or practical limits in field exploration budget—there are always some inherent uncertainties in heavy civil construction projects. On the other hand, the impacts of unanticipated conditions during construction can be addressed, to some extent, during design through alternate design concepts. As a minimum, the specifications should contain a set of criteria to define what constitutes a changed condition.

An example is given below of a post-tensioned rock anchor project in a competent granite bedrock foundation. Although all of the boreholes at the project site show the foundation is competent granite, the site geology for the project indicates that potentially soft decomposed diabase dikes could be encountered, even though the possibility is quite remote.

Unforeseen foundation conditions shall correspond to one or more of the following conditions in the foundation bedrock:

1. *A rock type other than granite is encountered, with mechanical properties that are significantly weaker than those encountered in the borings performed for this project. The unconfined compressive strengths of the weaker rock shall be no more than 25 percent of the average unconfined compressive strengths of the granite tested for this project.*

2. *A highly fractured zone more than 5-feet thick, with rock quality designation less than 10 percent, is encountered.*

3. *A shear zone more than 5-feet thick, with unconfined compressive strength of less than 1000 pounds per square inch is encountered.*

If these conditions are suspected in the foundation, the Contractor shall obtain NX-size cores of the foundation bedrock and perform laboratory tests to demonstrate to the Engineer that unforeseen conditions exist. If unforeseen conditions exist, then the Contractor will be compensated for the sampling and testing of the rock. If the data does not prove the existence of the unforeseen conditions, then the sampling and testing costs shall be the responsibility of the Contractor.

The Contractor shall be entitled to additional compensation for additional anchor materials and installation costs if unforeseen foundation conditions exist.

There are several elements in the example specifications above that are considered recommended practice for changed conditions:

- The criteria to define a changed condition are quantified to the extent possible. Parameters such as unconfined compressive strength, rock quality designation, and depths of soft zones are preferred over qualitative terms such as *significantly different,* or *as determined by the Engineer.*

- In terms of testing costs and compensation for the anchor installation in the event of an unforeseen foundation condition, this specification is fair to the contractor. This specification is also fair to the owner because it protects the owner from claims that arise from a liberal interpretation of what "unforeseen conditions" are.

- The design engineer has not ignored the remote possibility of an unforeseen foundation condition, and therefore has installed a mechanism in the contract documents for an orderly and contractually proper way to handle a surprise.

Plan Quantity Variation

Variations in plan quantities are all related, one way or another, to measurement and payment. All general conditions in construction contracts contain provisions to handle this issue. However, these provisions will be useless if the bid schedule is not set up to handle unanticipated variation in plan quantities. For example, if the exact length of an end-bearing pile is not known during design, then the pile should be paid for on a per-linear-foot basis. If the exact dimensions of a buried structure to be demolished are not known, then the demolition should not be paid as a lump-sum item. If a mixed soil and hard-rock excavation is anticipated, then the excavation should not be paid for as one single "Unclassified Excavation" item. In most cases, when a contractor knows that he or she will be paid for work that involves some unknown conditions and does not have to bear all of the risk involved, there will be a decreased likelihood of disputes and claims. In addition, bids will be more representative of the work required.

Acts of God

The contractor should be compensated for additional construction costs and delays associated with the so-called acts of God. The Engineers Joint Contract Documents Committee (1990) defined "Acts of God" as fires, floods, epidemics, and abnormal weather conditions. No contractor will be expected to allow for these unusual conditions in his or her bid. At the same time, the owner should not expect the engineer to include allowances in his or her estimate for these conditions. When these conditions occur, the owner would ultimately bear the additional cost. As design engineers, there are certain things that can be done during design to minimize the occurrence of "acts of God" during construction:

- Schedule the construction for certain times of the year during which damaging weather is less likely. For example, flooding along a stream bank worksite can be minimized by scheduling work in the dry fall season, and not during spring runoff.

- Provide adequate guidelines on what will be allowed under extreme weather conditions, such as cold weather placement of concrete and earthfill. Work should be shut down during unacceptable conditions.

The dividing line between what is considered normal conditions and "acts of God" could be a subject of disputes and litigation. It may be relatively simple to declare a 500-year flood as an act of God, but will a 50-year flood be considered the same?

15.11   Owner-Furnished Equipment and Materials

Occasionally, the contractor is required to install equipment and materials furnished by the owner. Some of the reasons for which an owner would furnish his or her own equipment and materials include:

- The project schedule is so tight that there is not enough time for the contractor to go through a normal equipment procurement process. For example, specially fabricated gates and valves may take several months for shop drawing preparation, approval, and fabrication, and these items are needed at the beginning of the project.

- The owner can obtain the materials at a lower cost than he or she could by purchasing them through the contractor, either through an existing purchase agreement with a local supplier, or merely through savings in contractor's markup.

When an owner furnishes his or her own equipment or material, he or she bears the responsibility for the quality of the equipment or material, and the contractor is only responsible for installation. When an owner designates his or her own borrow site for a fill, he or she is responsible for the material characteristics and adequate quantity required for completing the work. It is not fair to ask the contractor to comply with the material specifications of an owner-designated borrow site. Material specifications for owner-furnished equipment and materials should state explicitly so. There should be no contractor quality control testing, even though the owner may provide his or her own testing during installation and placement. All of the inherent properties and characteristics of the owner-furnished equipment and materials should still be included in the product specifications for the following reasons:

- The contractor needs this information for installation. Information that will be helpful to the contractor includes shop drawings from the manufacturer, product data sheets, and operation and maintenance manual. All of the design criteria for that piece of equipment should also be listed if they are critical for the installation and performance of other related work that is the responsibility of the contractor.

- The significant properties of earthfill material should be included because these properties would affect required equipment and how the fill will be placed to meet placement requirements. For example, if the fill is a clay, then a sheepsfoot roller or a tamping roller is more appropriate than a vibratory roller for compaction. The converse is true if the fill is a clean sand and gravel material.

15.12   Site-Safety Issues

This section pertains to the safety of the construction site, and not on safety issues of the end-users of new facilities. Although the design engineer often designs a new facility that meets all applicable industry safety guidelines, frequently the site considerations of constructing his or her design are often ignored in the design process. Construction site safety is one of the most important topics in construction management, but is beyond the scope of this book. An excellent discussion on construction site safety can be found in Fisk (1992). Because the contractor exercises direct control over his or her own equipment, workers, and means and methods of construction, construction safety is the sole responsibility of the contractor. It is his or her responsibility to comply with state and federal safety regulations, such as those stipulated by OSHA (1990). When an accident occurs during construction, the owner and the engineer are frequently involved in litigation to sort out who is liable for damages, regardless of whether the owner or engineer bears any part of the responsibility.

What can engineers do during design to address construction safety? It is suggested that the best time for designers to get involved with safety is before construction, during planning and design (Gambatese, 2000). Gambatese suggested the following practice during design to reduce the risk of construction safety:

- Decrease the amount of scheduled night work and overtime. On the other hand, it should be noted that restricting the contractor's schedule would take away some of the contractor's flexibility in performing the work, and may result in higher construction cost.

- List hazardous materials and note their location on the site.

- Locate underground utilities and other below-grade features.

- Design site grades to minimize the amount of work done on steep slopes.

- Incorporate safety issues as part of the constructibility review during design.

All of the suggestions listed above are excellent. The following elements should be included in the specifications regarding site safety:

- Compliance with OSHA construction regulations (1990)—Besides being legally obligated, the contractor is made contractually obligated to provide a safe working construction environment.

- Health and safety plan—Requiring the contractor to prepare and submit a health and safety plan forces him or her to address safety issues before they occur.

- Safety provisions—It is within the realm of the owner's rights to request certain minimum safety provisions in a construction site, such as warning signs, hard hats, steel toe boots, safety fences, first aid kit, fire extinguishers, and working telephones.

- Work-stoppage provisions—The contract should contain provisions for the field engineer to stop work for imminent hazards or dangerous conditions. Work stoppage always implies delays and claims, and should therefore be exercised with prudence and judgment by the field engineer. However, in the event of a serious or fatal injury, the financial consequences of a lack of action on the part of the owner's representative could be significantly higher than those associated with a temporary work shutdown to correct safety deficiencies.

CHAPTER 16

GOOD SPECIFICATION-WRITING PRACTICES

16.1 Literary Style

The writing style for technical specifications is interesting: the document is a legal document, yet it is written in a simple and brief style. Perhaps the reason is that technical specifications are directions for contractors, material suppliers, and product manufacturers. They are not written for literary scholars, and they are certainly not written for lawyers. Technical specifications are not expected to win any literary awards. In fact, most good contractor-preferred technical specifications contain few multisyllabic words, have short and brief sentences, and perhaps are considered boring by most standards. Yet, this simple, well-planned document allows contractors to understand exactly what materials they need and how to install them. Coincidentally, this is also the type of technical writing preferred by engineers.

The first requirement in technical specifications is to communicate the technical information to the builder. Significant technical and design issues pertaining to specifications are discussed in detail in Chapter 15. When a set of specifications are technically correct, the next responsibility of a specification writer is to convey this information quickly to the contractor, without ambiguity or delay. This chapter explains how some general guidelines can help to accomplish such communication. There will be no English or grammar lessons here. It is assumed that the specification writer has the basic writing skills and vocabulary that he or she needs for any technical writing. For more detailed guidelines on specification language (spelling, sentence structure, capitalization, punctuation, grammar, etc.) the reader is referred to an excellent treatment on this subject by the Construction Specifications Institute (CSI 1996f).

16.2 Recommended Guidelines

When a set of specifications are technically correct, the project probably can be built—even if the language is poorly written—but the engineer may need to clarify and explain the intended meanings during construction. When a set of specifications contain technical errors and design flaws, the most well-written language that follows every rule and guideline, can result in claims, disputes, and even litigation. Therefore, in the interest of giving the utmost attention to technical contents, the author de-emphasizes the importance of language that is considered good for specification writings. It is with this caution that the following guidelines are recommended to form the literary basis for an all-around well-written specification.

Imperative Mood

Instructions given to contractors are in the imperative mood and are written in one of two ways:

> *The Contractor shall coat the handrails in two 10-mil layers of brown paint.*
>
> *Coat the handrails in two 10-mil layers of brown paint.*

The two instructions are exactly the same, and are both acceptable. The latter is the preferred format of the CSI. In general, actions of the contractor are given by using *shall* (absolute), and the actions of the owner and engineer are given by using *will* (optional), as follows:

> *The engineer will perform field density testing of the in-place compacted fill. If the tests indicate that the compacted fill does not comply with the specifications, the Contractor shall rework failed materials until the specified density is met. Reworking may include removal, recompacting, or reconditioning, or combinations of these procedures.*

Avoid Repetition

There is a saying related to specification writing: *Say it once, and say it right.* Each requirement has its own logical place and should not be repeated elsewhere. All references to a particular requirement can be made by citing the appropriate article, paragraph, or subparagraph. For example, it is determined that a Type-II low-alkali cement is required for a project that has several different types of concrete—cast-in-place structural concrete, cast-in-place lean concrete backfill, precast concrete, and mortar grout. The cement specification is contained in paragraph 2.1 of Section 03310: Structural Concrete. The cement specification of the remaining concrete in other specification sections for this project is then referenced to paragraph 2.1 of Section 03310. Any necessary changes for the cement type can be made only in paragraph 2.1.

Use of Abbreviations and Symbols

Because of potential ambiguity and misunderstandings, the use of abbreviations and symbols is not considered good practice in specification writing. There is another saying related to specification writing: *When in doubt, spell it out.* Consider the following examples:

> Recommended: *Variation of the overall slope line from a straight line shall not exceed 12 inches in 100 feet, 3 inches in 30 feet, and 1 inch in 10 feet.*
>
> Not recommended: *Variation of the overall slope line from a straight line shall not exceed 12 in. in 100 ft., 3 in. in 30 ft., and 1 in. in 10 ft.*

As a minimum, if abbreviations are used in specifications, they should be clearly defined the first time they are used, as illustrated in the following example:

> *The 90-day compressive strength of the roller-compacted concrete shall be 2000 pounds per square inch (psi). Additional strength requirements shall be as follows:*
>
> | | |
> |---|---|
> | *7-day compressive strength:* | *500 psi* |
> | *28-day compressive strength:* | *1500 psi* |

Definitions of abbreviations should only apply to a particular section, and should be repeated in other sections when the same abbreviations are used. Also, a listing of abbreviations should be included on a drawing or a specification section.

Although abbreviations and symbols are routinely used to shorten callouts or labels in construction drawings, they are not recommended in specifications. When abbreviations and symbols are used in construction drawings, they are defined in the drawings. Use of symbols, with or without definitions, is discouraged in specification text. For example, the following symbols should be spelled out in their entirety:

Symbol	Spelled-out version
%	percent
°	degree
C	Centigrade
F	Fahrenheit
@	at
/	per
+	plus
−	minus

Abbreviations and symbols, however, are acceptable in tables and schedules in specifications with limited space.

Avoid Highlighting Text

Specification text should not be bold, underlined, italicized, or placed in quotations. Text that is highlighted in some way suggests to the contractor that it is more important than other parts of the specifications. The contractor should consider the entire specification equally important.

Avoid Irrelevant Text

A common practice in specification writing is to edit a specification section used in a previous project. There is nothing inherently wrong about this practice, except that irrelevant text and requirements should be removed to avoid confusion. For example, if the current project does not have fill density testing, then all references to fill testing standards should be deleted from the earthwork specification. If the current project does not have waterstops in the concrete joints, then all references to waterstops should be deleted from the concrete specification. Irrelevant text and requirements only confuse the contractor, and it should not be the contractor's responsibility to guess where these unnecessary requirements apply in the project.

Avoid Long, Blocky Text

Long, blocky text should be broken up into shorter, separate paragraphs for the following reasons:

- A long, blocky paragraph most likely contains more than one idea or requirement. Each idea or requirement deserves a separate paragraph.

- When a long blocky paragraph is being referenced from other sections or during construction, it is not immediately clear what portion of the requirement is being referenced without further explanation.

Consider the following blocky text that is common in some federal agencies' specifications:

C. *Excavation and bedding—When the surface of roadway embankment in which a pipe culvert will be placed has reached an elevation approximately one-fourth the diameter of the pipe above the prescribed elevation of the invert of the pipe, the embankment material shall be excavated carefully to the established lines and grades to provide a firm and uniform bearing for the entire length of the pipe. In original ground, the trenches for pipe shall be excavated to a bottom width equal to the diameter of the pipe plus 1 foot and to slopes of 1H:1V (horizontal:vertical). The trenches in which pipe will be laid shall be excavated carefully to the established lines and grades to provide a firm and uniform bearing for the entire length of the pipe. Where rock is encountered in the bottom of a trench, the trench shall be excavated to a depth of 6 inches below the grade established for the bottom of the pipe, and this additional excavation shall be backfilled with approved material which shall be tamped thoroughly in place before the pipe is laid. Where the character of the material at any point in the bottom of the trench is such as might cause unequal settlement or provide unequal bearing for the pipe, as determined by the Engineer, the unsuitable material shall be removed to such depth as may be directed and the additional excavation shall be backfilled with approved material which shall be tamped thoroughly to insure an even and unyielding foundation for the pipe.*

The above text can be improved by using shorter paragraphs, as follows:

3.3 EXCAVATION AND BEDDING

A. *When the surface of roadway embankment in which a pipe culvert will be placed has reached an elevation approximately one-fourth the diameter of the pipe above the prescribed elevation of the invert of the pipe, the embankment material shall be excavated carefully to the established lines and grades to provide a firm and uniform bearing for the entire length of the pipe.*

B. *In original ground, the trenches for pipe shall be excavated to a bottom width equal to the diameter of the pipe plus 1 foot and to slopes of 1H:1V (horizontal:vertical). The trenches in which pipe will be laid shall be excavated carefully to the established lines and grades to provide a firm and uniform bearing for the entire length of the pipe.*

C. *Where rock is encountered in the bottom of a trench, the trench shall be excavated to a depth of 6 inches below the grade established for the bottom of the pipe, and this additional excavation shall be backfilled with approved material which shall be tamped thoroughly in place before the pipe is laid.*

D. *Where the character of the material at any point in the bottom of a trench is such as might cause unequal settlement or provide unequal bearing for the pipe, as determined by the Engineer, the unsuitable material shall be removed to such depth as may be directed and the additional excavation shall be backfilled with approved material which shall be tamped thoroughly to insure an even and unyielding foundation for the pipe.*

Use Streamlined Format

The streamlined format is very effective for listing several requirements. It is important that every item that is listed is assigned a subparagraph number for referencing. An example of streamlining is shown as follows:

2.2   CONCRETE MATERIALS

A.   *Cement: ASTM C150, Type II, gray.*

B.   *Fine and coarse aggregates: ASTM C33.*

C.   *Water: Clean and potable and not detrimental to concrete.*

D.   *Compressive strength at 28 days: 3000 pounds per square inch.*

E.   *Maximum water:cement ratio: 0.45.*

F.   *Slump: 2 to 4 inches.*

G.   *Air entrainment: 4 to 6 percent*

CHAPTER 17

Types of Construction Specifications

17.1   General

There are many ways for an engineer to specify construction products, materials, testing, and installation. In general, the engineer chooses—based on one or more of the following considerations—the best method for a certain feature of the design:

- Based on the owner's requirements, the method will yield the best product and workmanship from a technical and performance standpoint.

- The method will encourage the best price competition.

- The method will incorporate the advantage of the contractor's experience, special expertise, and innovation.

Different types of specifications include *descriptive (or method) specifications, performance specifications, standard reference specifications,* and *proprietary specifications.* For a given set of project specifications, it is common for all four methods to be used to achieve the goals of the designer. In this chapter, each of these methods is described, and examples are given to illustrate each method and its advantages and disadvantages.

Many public agencies (e.g., county engineering departments, state transportation departments, and certain federal agencies) have their own standard specifications for the type of work these agencies typically perform. Because they are so similar to each other among the states, some of these *agency specifications* are described with an emphasis on standard transportation specifications. In addition, when a design firm is contracted to perform design work for these agencies, the agencies require that their own specifications be used, and certain procedures and protocol are used by the design firm to satisfy specific project requirements.

17.2   Descriptive Specifications

Descriptive specifications give the properties of a material, product, or equipment without mentioning a brand name or without specifying the end result. The performance or end result of the material, product, or equipment is the responsibility of the design engineer. When a contractor is asked to furnish something based on a descriptive specification, he or she is not responsible for the performance of that product or material, as long as it meets the specification requirement. Examples of descriptive specifications are:

Fill material—The fill shall be a mixture of clay, silt, sand, and gravel with a maximum particle size of 3 inches, a minimum of 20 percent passing the No. 200 sieve, a minimum plasticity index of 10, and a maximum liquid limit of 40. The fill shall be classified as either clayey sand (SC) or sandy clay (CL), based on the Unified Soil Classification System. The fill shall be free of peat and other organic materials, debris, and trash.

Mortar mix—The mortar mix shall consist of two parts ASTM C33 fine aggregate and one part ASTM C150 Type I cement, with enough water added to obtain a workable consistency.

Threaded fasteners—Structural threaded fasteners shall conform to ASTM A325-93. Connections shall be bearing types.

Metal pipe—Metal pipe shall be round corrugated galvanized steel pipe, 18 inches nominal diameter conforming to the requirements of ASSHTO M36-90, minimum 16 gauge.

Fill compaction—Compact the fill with at least two passes of a vibratory plate compactor with a minimum dynamic force of 3000 pounds. No testing of the fill is required.

When an engineer uses descriptive specifications for a material or product, he or she is reasonably confident that the end result or design intent will be obtained. In the examples given above, if the contractor furnishes and places the specified fill material, the in-place material should provide the minimum strength and water-tightness that are required in the design. Or, when the strength of the mortar specified above is not critical, then—provided the contractor achieves the consistency that is needed for workability—the mortar mix is considered adequate.

Because the contractor is not liable for the performance of the material or product, assurance of the end result should not be the responsibility of the contractor. For example, if the contractor compacts the fill with the required two passes of a piece of approved equipment, then he or she is not responsible for the density of the fill. The engineer may still want to test the density of the fill after compaction for record purpose, but not for compliance with the specifications. If the density of the fill is inadequate after two passes, the engineer can issue a field order to increase the number of passes, with an associated increase in payment to the contractor for increased effort.

A descriptive specification is straightforward and requires little risk on the part of the contractor. Because of low risk, the prices for items based on descriptive specifications are generally very competitive and consistent among all of the bidders. In this sense, use of descriptive specifications should be encouraged as much as possible. This method uses the technical know-how and experience of the engineer. Because of the higher risk imposed on the engineer, inexperienced design engineers should be careful in using descriptive specifications, especially when the descriptive specifications are simply copied from a previous project.

17.3   Performance Specifications

A performance specification gives the end result or end product without mentioning the means and method to achieve the requirement. When this method is used, the burden of performance is on the contractor, and therefore, the contractor assumes a significantly higher risk. There are several situations in which performance specifications are appropriate:

- This method encourages the use of contractor's innovation, experience, and technical know-how, and may result in an improved product.

- This method allows the contractor to use the means and method most cost-effective for the contractor, and may result in a cost saving to the project.

- Certain construction products and methods are so specialized that it is not possible for an engineer to use descriptive specifications at all.

Examples of performance specifications are:

Concrete strength—The specified 28-day compressive strength of concrete shall be a minimum 4000 pounds per square inch (psi). Concrete materials shall consist of ASTM C150 Type I, ASTM C33 coarse aggregate size 67, ASTM C33 fine aggregate, 2-inch to 4-inch slump, and a maximum water-cement ratio of 0.45.

The engineer specifies the quality of the materials, but leaves the concrete mix design to the contractor, who is responsible for proportioning the batch weights of the concrete materials and testing the concrete cylinders to verify specified strength. This method is very cost-effective for ready-mix concrete.

Percent compaction—The fill shall be compacted in maximum 12-inch loose lifts to at least 95 percent of the maximum dry density in accordance with ASTM D698-78 within plus or minus 2 percent of the optimum water content.

The contractor determines the type of compaction equipment and the number of passes required to achieve the specified percent compaction. Field testing of the in-place fill by the engineer is essential to verify compliance with specifications.

Anchor design strength—The contractor shall determine the bonded length of the anchor and diameter of the anchor drill hole to achieve a design strength of 100 tons. The anchor materials shall meet the requirements of paragraph 2.3.A and applicable standards of the Post-Tensioning Institute.

Based on the foundation conditions given in the construction documents, the contractor is required to furnish 100-ton anchors by sizing the drilled holes, the bonded length, and the method of installation. Verification of design strength occurs through a series of tests for anchors that may include performance testing, proof testing, and lift-off testing. Industry practice for anchor materials and installation is determined by the Post-Tensioning Institute. Unless it can be proved that there is a change in foundation conditions, the contractor is responsible for correcting problems associated with failed anchors.

Precast concrete box culvert—Precast concrete box culverts shall be reinforced concrete conforming to ASTM C789-87b, 4-foot span and 3-foot rise internal dimensions, and with a wall thickness of 5 inches. The minimum 28-day compressive strength shall be 4000 pounds per square inch. Precast concrete sections shall be designed for a minimum earth load of 30 pounds per square foot, and a hydrostatic load of 10 feet of water.

The contractor's responsibility here is to design the reinforcements for the culverts to meet standard ASTM specifications for the specified loads. Verification of compliance is usually handled through submittal of shop drawings and certification from the precast concrete supplier concerning the material types and strength.

A set of performance specifications shall contain three main components:

1. Requirements—The performance, end product, or end result should be expressed in properties that can be quantified and verified by testing. In the example of the concrete mix, the requirement is the 4000 psi strength. In the example of the anchors, the requirement is the 100-ton design load.

2. Criteria—One or more design criteria the contractor is required to comply. In the example of the concrete mix, the criteria are the water-cement ratio and slump. In the example of the precast concrete, the criteria are the soil load and hydrostatic load. In the example of soil compaction, the criterion is the percent compaction in accordance with ASTM D689 compaction standard.

3. Testing—Verification of performance is obtained through testing. Testing can be performed by the engineer, the contractor, or both.

Requirements and criteria should be achievable based on project site conditions and current technology and construction methods. If the engineer is uncertain as to whether the required performance can be achieved, he or she should consider performing testing during the design phase to obtain design parameters prior to construction, and the test information should be furnished to the bidders during bidding.

17.4   Standard Reference Specifications

Reference specifications are publications that contain standards of material and product quality, design standards, quality of workmanship and installation, test methods, and codes. This information is used to represent standard of quality, accepted methodology, uniformity, and minimum standards. When a standard reference specification is used in a project specification, that standard reference constitutes part of the construction document as if it is included in whole in the document.

Using standard reference specifications offers many advantages for establishing minimum quality, bid uniformity, and uniformity of construction testing and acceptance. It also reduces the need for the engineer to prepare a lengthy text on the technical requirements.

The following is a partial list of reference standard agencies that may be of interest to civil designers:

American Society for Testing and Materials (ASTM). The ASTM provides the most widely used standards for materials and test methods of products that include iron and steel, alloys, concrete, soil and rock, geotextiles, plastics, rubber, etc.

American Concrete Institute (ACI). This organization publishes standards on the properties and application of concrete. Essentially all concrete placement specifications and workmanship issues can be addressed using ACI standards.

American Institute of Steel Construction (AISC). This organization is a nonprofit, technical-specifying, and trade organization for the fabricated steel industry in the United States. AISC provides specifications and codes for the design of fabricated steel.

American Association of State Highway and Transportation Officials (AASHTO). This organization publishes standards on highway materials specifications and testing methods. Some of the AASHTO standards and ASTM standards are identical.

American Water Works Association (AWWA). This organization publishes standards on gates, valves, and pipes for the application of water control.

A standard reference specification requires the following designations:

- Agency

- Number of the standard and year of issue

- Title of the standard

The following are some examples of reference standards:

ASTM D698-91: Test Method for Laboratory Compaction Characteristics Using Standard Effort.

ASTM C150-80: Specification for Portland Cement.

ACI 306.1-90: Standard Specification for Cold Weather Concreting.

It is important that the engineer who employs a standard reference specification is familiar with that specification. If the engineer is unfamiliar with the specification, serious problems may arise:

- Duplication or contradiction within the construction documents—Requirements in standard reference specifications should be checked for any duplication or contradiction within the requirements in the drawings and technical specifications. For example, if the design loads for a precast concrete culvert (ASTM C850) are specified, that project loading criteria should be checked against the standard design loads used in ASTM C850, which uses either HS20 or Interstate Loading.

- Hidden choices—The reference specification may contain a number of choices of materials and products that are covered under that designation. If a particular type of material or product is not specified, the selection of that material is uncertain, and what takes place beyond this point is up to the contractor. The contractor may request additional information, or he or she may simply assume the cheapest one on the list during bidding, and then be compensated for additional costs for the more expensive item originally intended by the designer. Examples of reference specifications with hidden choices include ASTM C150 for the types of cement, ASTM C33 for the size of the coarse aggregate, and ASTM C494 for the type of chemical admixture.

- Unavailable product—The reference specification contains a list of materials or products that are manufactured to comply with the requirements of that standard. If the specified properties fall outside of the list of products that are manufactured, specifying a product to meet that standard does not necessarily guarantee that this product will be available. It is not reasonable to assume that the manufacturer will specially fabricate that product solely for one project, especially in small quantities. When such a situation occurs, the contractor may request a substitution or an equivalent product, which may be more expensive and inferior to the originally specified product. As an example, a designer specifies a 12-inch PVC pipe with a standard dimension ratio (SDR) of 23.5, meeting ASTM D3034-89. Without reviewing the pipe dimensions table in this specification, the designer is unaware that SDR 23.5

for this type of pipe is available only for 4-inch and 6-inch diameters, but not for 12-inch diameter.

It is clear from the discussion above that, as a general rule, the designer should carefully read all of the contents of a reference specification before adding it to his or her project specification. A hard copy of each reference specification should be included in the project file during design and at the project site during construction. Updates of the reference specification issued after award of the contract have no relevance to the contractor, and new or changed details cannot be required of the contractor.

17.5 Proprietary Specifications

A proprietary specification specifically calls out a brand name, model number, manufacturer of a product, or a product or method of construction that is patented. Proprietary specifications are used for the following reasons:

- The design engineer is familiar with the good performance of certain products, either from past experience or from reported literature.

- The named product is compatible with an existing project component that is being rehabilitated or replaced.

- The owner prefers the named product over other products.

- No generic product is available to satisfy the technical intent for a particular project.

- The brand name or manufacturer is called out to assist the contractor in identifying the source or sources of his or her suppliers, especially for very specialized items that are relatively new to the market.

Examples of proprietary specifications are:

Mechanical gate—Contractor shall furnish and install one cast-iron sluice gate, 48-inch by 48-inch nominal dimensions, with hydraulic operator control system. The gate materials shall conform to the requirements of paragraph 2.3.A. Acceptable gate manufacturers include Rodney Hunt, Hydro Gate, and Waterman, or approved equal.

Concrete revetment system—Contractor shall furnish and install Tri-Lock Model Number 4015, 6-inch nominal thick precast concrete blocks, supplied by the American Excelsior Company, or approved equal.

Coatings—The hydraulic pumping unit base, tubing, and accessories shall be painted as follows:

1. *Surface Preparation: SSPC-SP-6.*

2. *First Coat: Epoxy Coating (Amercoat 370), 5.0 mils thick.*

3. *Second Coat: Epoxy Coating (Amercoat 370), 5.0 mils thick.*

4. *Third Coat: Aliphatic Polyurethane (Amercoat 450HS), 2.0 mils thick.*

The most restrictive type of proprietary specification specifies a single product or manufacturer and allows no substitutions. This method allows the engineer to have direct control over the specified product, but it also represents the least-price competitive method.

Therefore, from a cost standpoint, this method should be used only when there is no other choice available in the market, or a very particular need of the owner is mandated.

It is common for an engineer to specify a proprietary product with more than one available source. This allows the contractor to substitute a similar product that is not listed in the specifications. This method gives the engineer some assurance of the quality of the product and offers the advantage of price competition. When this method is used, a list of product properties and criteria should be added to the specification for two reasons:

1. They allow the contractor to shop for the least expensive product meeting these requirements.

2. They form the contractual basis for the engineer to review a substitution for approval or rejection. To avoid disputes, the contract should state that the engineer's approval or rejection of a product substitution is final and is not subject to appeal.

Most proprietary specifications are based on guide specifications furnished by the product manufacturer. Unless a restricted proprietary specification is intended with no substitution allowed, the manufacturer's guide specifications should not be used verbatim. If open competition is allowed for the proprietary products, then the engineer should review the guide specifications carefully to remove any provisions that will restrict other manufacturers from competing on this product. For example, the technology of cure-in-place pipe liners is a proprietary process used in repairing defects in underground pipes. There are several of these systems available in the United States, and all of them are somewhat similar, but the curing temperatures of the liner resin for each differ. If a proprietary specification for pipe repair is based on a particular guide specification, the curing temperature that is unique to that particular system may prevent other competitors from bidding on the project. In this case, the designer should research the range of curing temperatures and specify an acceptable range. A better way to handle the different curing temperatures is to use the phrase "as required by the liner supplier."

17.6 Agency Specifications

Some county, state, and federal agencies have their own standard technical specifications for construction. These agencies include county engineering departments, state transportation departments, and some federal agencies, such as the Federal Highway Administration, Natural Resources Conservation Service (formerly the Soil Conservation Service), and the U.S. Army Corps of Engineers. The technical content of these specifications depends on the type of work these agencies normally do, and is too diverse to list here. By far, the most prevalent and widely used agency specifications are the state highway department specifications.

Every state in the United States has its own standard highway specifications. The names of these specifications vary somewhat. For example, they are called *Standard Specifications for Road and Bridge Construction* in Illinois and Colorado, and *Standard Specifications for Highway Construction* in Oregon and Oklahoma. For simplicity, a generic name of *standard highway specifications* is used here for discussion. Regardless of the title, all of these standard highway specifications are structured similarly to the federal highway specifications, and contain work requirements related to highway construction and maintenance:

- Earthwork, landscaping, and erosion control

- Subgrade, subbase, and base course

- Bituminous surfaces and pavements, concrete pavements and sidewalks, pavement rehabilitation, and shoulders

- Bridges, culverts, sewers

- Drainage, lighting, and safety provisions

- Materials and equipment specifications

These standard highway specifications are used mainly for highway construction and maintenance work by each state's highway department. However, the use of these standard specifications goes far beyond individual highway departments:

- Local sand and gravel manufacturers use these specifications to make concrete aggregates, base courses, riprap, and other materials.

- The local paving contractors use these specifications for placing unpaved roads, asphalt pavements, and concrete pavements.

Because local material suppliers and contractors follow state highway specifications, it is logical for an engineer to specify materials and placement procedures based on state highway specifications. Indeed, this practice has been widely used in the heavy civil construction industry, regardless of whether the work has been intended for private-sector, municipal, county, state, or federal agencies. It is not surprising that state highway specifications are frequently used in heavy civil construction in much the same way that standard reference specifications (see Section 17.4) are used. When state highway specifications are used for a project, these specifications are made a part of the project specifications. State highway specification provisions have the following advantages:

- The specified standard materials will most likely be available locally and do not need to be specially manufactured.

- The local contractors are familiar with the requirements, and thus will be more likely to be proficient in that work.

- The pricing will be more competitive.

Examples of standard highway specifications being used in project specifications are:

Aggregate base course—Aggregate base course shall conform to Standard Specifications for Road and Bridge Construction, Colorado Department of Transportation, 1991, Section 703.03: Aggregates for Bases, Class 4 Aggregate Base.

Bituminous pavement materials—The asphalt cement shall be viscosity grade AC-20 in accordance with Massachusetts Department of Public Works Highway Specifications, 1988, Designation M3.11.06: Bituminous Materials.

Riprap bedding—Riprap bedding shall be free-draining, sound and durable cobbles and gravels or manufactured crushed rock meeting the soundness and durability requirements of

Colorado State Department of Transportation, Standard Specifications for Road and Bridge Construction, 1991, Section 703.09, Filter Material, Class A.

17.7 Considerations for Federal Projects

A number of issues should be considered when an engineering firm is engaged by a federal agency to prepare construction documents. Issues concerning fee schedules, design criteria, contract administration, and submittal review, though all important, are outside the scope of this book. The following is a discussion of issues concerning preparation of technical specifications:

- All federal construction is contracted using Federal Acquisition Regulations (FAR), which represent all front-end documents for bidding and contracting. These FAR documents are similar to the bidding documents, General Conditions, and Supplemental Conditions in the private sector. The engineering firm is asked to prepare the technical specifications, but the front-end documents are prepared by government personnel (specifically, the contracting officer). It is important that there is coordination between FAR requirements and technical requirements, especially for Division 1 (General Requirements). To accomplish this goal, an engineer can review all applicable FAR clauses that will be used, or the government agency can review all technical specifications for coordination with the FAR requirements. The latter approach is preferred because most engineers are unfamiliar with FAR.

- Use of restricted proprietary specifications is generally not allowed when there is no competition for a particular product. The reasoning for this requirement is to encourage free competition in federal procurement. That does not mean that proprietary specifications cannot be used. When proprietary specifications are used, the named product should be accompanied by a list of basic properties and criteria by which equivalent or equal substitutions can be evaluated.

- When a specified material or product is made outside of the United States, a similar material or product should be available in the United States. This is a direct requirement from the *Buy American Act.* Specifying a foreign product without an American counterpart is not allowed, except under special circumstances in which domestic products cannot meet the required project criteria.

- The names of the parties involved in construction and frequently used in specifications (the owner, engineer, and contractor) are somewhat different. The term *contractor* remains unchanged. The term *owner* becomes *Government,* and the term *engineer* becomes either *contracting officer* or *contracting officer's representative,* depending on the circumstances. For matters of administration and contract-related issues, such as measurement and payment, impact to schedule, or other contract changes, the term *contracting officer* should be used. For matters related to technical issues, submittal review and approval, field inspection and approval, the term *contracting officer's representative* should be used.

CHAPTER 18

Construction Specifications Institute Format

18.1   Introduction

The Construction Specifications Institute (CSI), in association with Construction Specifications Canada, is a leader in North America in developing and updating guidelines to the structure and formats of construction documents. The CSI format for technical specifications has been widely adopted in the architectural engineering and construction professions in the United States, including some state and federal agencies.

In this chapter, the CSI format for technical specifications is introduced and discussed, with emphasis on heavy civil construction projects. When a set of project specifications is written in CSI format, it implies that the recommended formats are followed. Historically, the CSI is deeply rooted with the building construction industry, and understandably, most of the CSI format and guidelines are for architectural construction. Because this book is intended for civil engineers and deals with civil construction projects, architectural elements of the CSI are deliberately de-emphasized, and only elements related to heavy civil construction are highlighted.

If the reader is interested in a historical development of specification formats, a good reference can be found in *Construction Specifications: Writing, Principles and Procedures* (Rosen 1999).

18.2   MasterFormat

The CSI MasterFormat (CSI, 1995) is a system for organizing detailed construction information into a standard order or sequence based on products and construction methods. In this system, all construction products or methods are organized under 16 divisions (CSI, 1996a and 1996b):

Division 1—General Requirements

Division 2—Site Construction

Division 3—Concrete

Division 4—Masonry

Division 5—Metals

Division 6—Wood and Plastics

Division 7—Thermal and Moisture Protection

Division 8—Doors and Windows

Division 9—Finishes

Division 10—Specialties

Division 11—Equipment

Division 12—Furnishings

Division 13—Special Construction

Division 14—Conveying Systems

Division 15—Mechanical

Division 16—Electrical

The subject matter and types of construction for the divisions are self-explanatory for some divisions (e.g., Divisions 3, 4, 8, and 16), but are not so obvious for others (e.g., Divisions 10, 11, and 13). It is clear based on the division titles that the master format is organized for the building industry. However, all of the construction work from civil engineering construction can be logically fit into this format.

Each division is divided into sections. CSI has specific guidelines and recommendations for the subject matter, numbering system, and format for the sections. Each section has a five-digit number and a title. The first two digits of the section number (ranging from 01 to 16) refer to the division number and, therefore, are fixed for a given division. The third and fourth digits are "level 2" and "level 3" numbers, respectively, that are assigned by CSI. The last digit is the "level 4" number, which is available for users for specific project section numbers. This numbering system also contains gaps that can be used by users for nonstandard titles that are not on the CSI MasterFormat list (CSI, 1995).

Based on the section titles, the following is a discussion regarding the application of each division to civil construction work:

Division 1—General Requirements

This division contains general and administrative requirements, procedural requirements (submittals, quality control, safety, measurement, payment issues, etc.), and temporary facilities and utilities. All of these requirements are stated once in this division, and they apply to Divisions 2 through 16. The CSI indicates that the authority of Division 1 in governing the work for other divisions is not explicitly stated in most general conditions (including those of EJCDC) and should be explicitly established in the supplemental conditions (CSI, 1996c). Division 1 covers in more specific detail several topics (e.g., safety, submittals, testing) that are already covered in general conditions. Provisions in Division 1 should be an extension of the general and supplemental conditions, but should not be a repetition or contradiction to those conditions.

Division 2—Site Construction

All of the work items in this division are applicable for civil engineering construction. This division includes excavation, earthwork, site demolition, clearing, dewatering, drainage, tunneling, foundations, roadways, pavements, and reclamation. For some heavy civil construction, such as earth dams, tunnels, landfill liners and caps, and highway work, the majority of the work scope is covered in this division.

Division 3—Concrete

All of the work items in this division are applicable for civil engineering construction, including concrete foundations, bridges, spillways, outlet works, tunnel linings, pipes, and manholes. This division includes cast-in-place concrete, precast concrete, grouts, and mass concrete work. A relatively new use of concrete technology is the roller-compacted concrete (RCC) which has been used world wide in the application of water resources and pavements (Hansen 1991). Another construction method similar to RCC is soil-cement, which has been used in slope, bank protection, and soil stabilization for many years.

Divisions 4–16

Most of these divisions are for buildings, and only select portions of some of the divisions are commonly used for civil engineering construction. It is recognized that some civil engineering projects contain an architectural component and, therefore, require most of the Division 4 through 16 specifications. The reader is referred to CSI's MasterFormat for a complete list of section numbers in these divisions.

The MasterFormat is intended as a guideline only. Depending on the size and nature of a project, individual subscribers to this format exercise considerable freedom to choose the applicable divisions and section numbers. There are cases in which the selection of a division number can be left to the individual users. For example, a soil-bentonite slurry wall construction can logically fit into either Division 2 or Division 13. A flexible, prefabricated concrete revetment system can logically fit into either Division 3 or Division 13. At the same time, the selection of division and section numbers is kept within the general framework of the CSI MasterFormat.

When a division is not used in a project specification set, that division is still shown on the Table of Contents, but the phrase "Not Used" is included.

18.3   SectionFormat

The standard CSI format (CSI, 1997) for a section consists of three parts:

Part 1—General

Part 2—Products

Part 3—Execution

Part 1 contains administrative and procedural requirements specific to a section. Part 2 contains requirements for materials, products, equipment, or fabrication. Part 3 contains installation, placement, and testing requirements.

This three-part section format is a revision of a previous version, which contained an additional part, *Part 4—Measurement And Payment.* Even though the current standard contains three parts, some owners still prepare specifications in four parts.

The CSI section format is a guideline of where information should be shown. How the information is presented in each of three parts (articles, paragraphs, subparagraphs, numbering, margins, capitalization, etc.) is contained in PageFormat, which is discussed in Section 18.4.

The following is a discussion of the items recommended in a CSI specification section. These items represent the Articles that are major headings under particular Parts. Article numbers should be used in particular sections, but are not used in this discussion.

Part 1—General

Section Includes—This is an outline of the scope of this section. The scope should only contain major work items, but not incidental work (preparation, coordination with other work, cleanup, etc.). The outline should be brief, and each item on the list should be limited to one sentence or one phrase if possible.

Products Supplied But Not Installed—This is a list of products or materials that are specified in Part 2 of this section, and the installation or placement specifications are not shown. If the products are installed for the project, then the section number containing the installation specifications should be referenced here. Otherwise, no other reference is needed.

Products Installed But Not Supplied—This article lists the products that are installed with specifications in Part 3 of this section, and using products specified elsewhere or furnished by owner. If the products are specified in other sections, those sections should be referenced here.

Related Sections—This is a list of sections that contain work directly related to the work in this section. This listing should contain the section number and the complete section title.

References—This is a list of standard reference specifications that are used in this section. With this reference list in Part 1, any reference to these standard specifications can be abbreviated, which shortens the length of the specifications. This reference list should contain only those specifications actually used in this section, rather than a complete "shopping list."

Definitions—This article contains definitions of special or unusual terms. Examples of terms that can be defined here are *passes, percent compaction, rock excavation, unclassified excavation, refusal,* and *well points.*

Design Criteria—When the contractor is required to design a particular feature of a project (e.g., a performance specification; see Section 17.3), the performance and design criteria are listed here.

Submittals—The submittal requirements are listed here. Submittals can include materials data and samples, product data, shop drawings, proposed method and sequence, etc. The procedure of the submittal process should be referenced to the submittal requirements in Division 1. When a submittal is required from the contractor, the following minimum information should be specified explicitly, regardless of the nature of the submittal:

* Title of the submittal (e.g., *Dewatering Plan, Concrete Mix*).

* Schedule of submittal related to some part of the work (e.g., "sediment and erosion control plan should be submitted 21 calendar days prior to any excavation work"). Calendar days should be used instead of working days or just days.

* Detailed list of submittal contents.

Testing Responsibilities—The article clearly defines the testing responsibilities of the contractor and the owner. The procedure of the testing process should be referenced to the quality control requirements in Division 1. Detailed testing requirements should be located in Part 3. Some specifications do not contain owner testing requirements because they do not concern the contractor. Adding that information is preferred for the following reasons:

- It allows the contractor to anticipate what tests will be performed by others so that proper and timely coordination can be arranged in advance.

- The design engineer preparing the specifications may not be involved during construction.

- Responsibility of retests can be defined.

- Procedure to correct failed tests can be defined.

Restrictions—This article alerts the contractor of the restrictions that could affect his or her work. Examples of project restrictions include schedule, traffic, working room, and noise. Reference to drawings may be required.

Protection—Existing or new facilities that should be protected should be listed here. Protected items may include trees, wetlands, cultural resources, threatened or endangered species, streams, buried utilities, adjacent properties, etc. Typical specifications require the contractor to be responsible for any damage to protected items.

Preparation/Sequence—This article contains a list of preparatory work that should be in place prior to performing work for this section. Care should be taken not to use this article to limit the contractor's means and method. Rather, the purpose of this article is to convey the designer's intent on a workable and constructible sequence. The contractor should be allowed to use a different sequence or preparation procedure that will accomplish the same result.

Part 2—Products

Materials/Products—More than one article may be required to list the materials or products specifications, with one article used to list each material. When a material or product is specified here, the material or product name should match exactly the name used on the drawings and on the bid schedule.

Equipment—More than one article may be required to list the equipment, using one article per piece of equipment. A descriptive specification, performance specification, or proprietary specification (see Chapter 17) can be used to describe the equipment.

Mixes—Materials formed by mixing more than one ingredient in a plant or in the field are considered manufactured products. Proportions, ingredients, and procedures are specified here. Mixes commonly used in civil engineering construction include structural concrete, bituminous concrete, mortar, grout, and soil-bentonite.

Fabricated Products—Fabricated products require both written and graphical specifications for complete description, and they generally require submittals for material compliance and shop drawings. Written specifications include material requirements, design criteria, performance criteria, and workmanship requirements. Graphical fabrication requirements are shown on the drawings, and should be referenced accordingly. Workmanship requirements should be located in Part 3. The remaining written specifications are located here, in Part 2. Commonly used fabricated products in civil engineering construction include reinforcing steel for structural concrete and fabricated metalwork (e.g., trashrack, ladder, grating, vent pipes, handrails, gates).

Part 3—Execution

Placement/Installation—The placement or installation requirements and procedures for the specified materials, products, equipment, mixes, and fabricated products are included here. For an orderly presentation and quick location, it is recommended that the procedures follow the same order in which the materials or products are listed in Part 2. Placement or installation procedures for material or products that are specified elsewhere are also located here.

Fabrication Requirements—This article contains workmanship requirements for specified fabricated products. For fabricated metalwork, these requirements pertain to qualifications of welders, welding standards, and special project requirements.

Testing Requirements—Field testing to verify compliance with design is stated here. Testing requirements include procedure, reference testing standards, data reporting, and acceptance criteria.

Tolerances—This article contains tolerance requirements that include thicknesses, deviation from design lines and grades, and angular deviation. Different tolerances are needed for different materials and construction methods.

Not all sections require all three parts. As a minimum, Part 1 should always be used. When Part 2 or Part 3 is not needed for a section, the part title is still listed with the phrase "Not Used" under it.

18.4   PageFormat

The guidelines for arrangement of text on each page of the specification section are shown in the CSI PageFormat (CSI 1999). The guidelines include margins, page arrangement, section headings and endings, page footers, designations of articles, paragraphs, and subparagraphs. When these guidelines are followed, the text is presented clearly and facilitates easy reading.

The most unique feature of the CSI PageFormat is the organization in which information is presented and referenced. Text information is organized as follows:

SECTION (five-digit section number plus section title)

PARTS (three parts with standard part numbers and titles)

ARTICLES (major subject titles with article numbers)

Paragraph (article requirements with letter designation)

Subparagraph (subordinate to paragraph with numbering)

The numbering system is important because it is used for referencing texts within the specifications. An article number includes a prefix for the part number, a decimal point, and one or two digits beginning with "1." The article title should be capitalized. An example of an article number is:

3.1 PLACING STRUCTURAL FILL

A paragraph number consists of an article number plus an upper-case letter, starting with "A." A paragraph may or may not require a heading, such as:

3.1 PLACING STRUCTURAL FILL

A. *Place structural fill in maximum 9-inch loose lifts and compact to at least 98 percent of the maximum dry density in accordance with ASTM D698-78 within plus or minus 2 percent optimum water content.*

The full paragraph designation for the example above is paragraph 3.1.A.

A subparagraph is used to elaborate on further requirements under a paragraph. Further subdivisions under subparagraphs are also subparagraphs. A subparagraph number consists of an article number plus a letter designation plus a number starting with "1." A subparagraph may or may not require a heading, such as:

2.4 CONCRETE MIX

A. *Provide concrete mix with the following characteristics:*

    1. Minimum 28-day compressive strength: 4000 psi

    2. Slump: 2 to 4 inches

    3. Maximum water/cement ratio: 0.45

Three subparagraphs are shown in the example above, with designations of 2.4.A.1 for the compressive strength, 2.4.A.2 for slump, and 2.4.A.3 for water/cement ratio. Further subdivisions are allowed by CSI, but are not recommended. Further subdivisions of a subparagraph (which are also called subparagraphs) can use alternating lower-case lettering and numbers, such as 2.4.A.1.a or 2.4.A.1.a.1, etc. Excessive use of subdivisions beyond the first subparagraph level indicates excessive paragraph length and would require additional articles or paragraphs. In that case, the lengthy article or paragraph should be broken up into shorter articles or paragraphs.

CSI PageFormat also contains guidelines for section headings, endings, and page footers:

Section heading and ending: The first page of each section should have the section number and title centered at the top (CSI 1999). The section title should be capitalized. The end of the section should be designated with the "END OF SECTION." The author prefers to use "END OF SECTION" followed by the section number. No further text is allowed beyond this designation. Any tables, schedules, forms, etc., should precede the page containing this designation.

Page footer: The minimum information in the page footer should include the section title and page number. The page number starts with "1," and contains a prefix, which is the section number. Other optional information includes the project name, date of the specifications, and the electronic file name.

18.5 Summary

This chapter contains CSI's guidelines for preparation of technical specifications. These guidelines are introduced with application to typical civil engineering projects. More specific guidelines can be found in publications available at the CSI (see List of Resources).

Although the format was conceived based on the building industry and architectural work, the CSI format can be used effectively for all civil engineering projects. The CSI format of 16 divisions, a three-part section, and page organization represents a promising trend in the direction of preparing specifications in a unified manner.

CHAPTER 19

MEASUREMENT AND PAYMENT PROVISIONS

19.1   Importance of Payment Provisions

As discussed throughout this book, particularly in Chapters 3 and 15, civil engineering design projects inherently contain many conditions that are not apparent until construction is under way. Most of these conditions are related to excavation, backfill, foundations, groundwater, buried structures, and other site work that is common to all heavy civil construction. Associated with these conditions is the risk to be taken by both the owner and the contractor. Section 15.10 describes various ways to handle unknown conditions so that the associated risk is shared appropriately between the owner and contractor; this chapter discusses how this can be accomplished through financial compensation to the contractor. It is fair to say that most construction disputes and claims are directly related to costs incurred by the owner and compensation to the contractor. When an owner, understanding the risk involved, feels that he or she is not overpaying the contractor for work within the contract scope, that owner is less likely to withhold payment to the contractor. When a contractor understands that he or she will be compensated for risk taken for work within or beyond the contract scope, he or she is less likely to file claims.

In heavy civil construction projects, the contractual mechanism to handle properly unknown conditions and risk is through the bid schedule and the associated measurement and payment provisions. The role and importance of the bid schedule and measurement and payment provisions frequently are overlooked by both owners and designers, yet these matters are the grounds for most construction disputes and claims. The bid schedule and measurement and payment provisions define how the contractor will be compensated. To account for a variety of field conditions and risk factors, most heavy civil construction projects—unlike building construction projects—are bid with a combination of unit-price items and lump-sum items. Determining what work should be paid as unit-price items and what work should be paid as lump-sum items requires careful consideration of the risk involved, unknown conditions, fairness to all parties involved, and judgment. This chapter contains a discussion of preparation of bid schedules containing unit-price items and lump-sum items, methods of measurements for unit-price items, payment of lump-sum items, provisions for changed conditions and/or excessive changes from bid schedule quantities, and common sources of payment disputes and claims.

19.2   Formulation of a Bid Schedule

A bid schedule is a breakdown of work items that the contractor uses when submitting his or her prices during bidding. The bid prices then become the basis for compensating the contractor for actual work during construction. Each item on the bid schedule contains a measurement and payment provision. A bid schedule is incomplete without the associated measurement and payment provisions that are typically included in Division 1 of the CSI technical specifications (see Chapter 18). For the following reasons, the proper time to prepare a bid schedule in a final design project is at the beginning of the design:

- The bid schedule contains the work items that will be used in construction drawings and technical specifications. The terminology used in the bid schedule also should be used in drawings and specifications. Establishing the bid schedule at the beginning of design defines most of the work items for the design team, improves coordination, and minimizes subsequent changes.

- The bid schedule identifies the outline of the required specification divisions and sections. This allows the specification writer to initiate the specification writing process early in the project.

- The bid schedule represents a basis for discussion with the owner regarding various parties' shared risk and unknown conditions associated with each work item. Because it defines the philosophy and basis of the design, such dialog between the owner and the design engineer should be started at the beginning of the project, rather than near completion.

Table 19-1 is an example of a bid schedule for a dam rehabilitation project. Some of the items in this example will be used for discussion in this chapter. The example also illustrates some key elements of a bid schedule:

Item Number

The item number is important for referencing with measurement and payment provisions, and also for quick referencing during evaluation of the contractor's progress payment requests during construction. The order of the items in the schedule is not important; however, there are two common ways to list the items:

- In approximate chronological order the work will be performed
- In order of increasing CSI section numbers

In the example shown, the first item is mobilization and demobilization, and the last item is seeding for reclaiming the disturbed areas at the end of the project.

Item Description

The description should be chosen carefully and should be as short as possible. The item descriptions in the measurement and payment provisions in the specifications should match verbatim. In addition, the material-related terms used in the item description (e.g., "filter/drain sand in Item 10," or "Type I Riprap in Item 13" in Table 19-1) should be exactly the same as those used in the construction drawings and technical specifications. The item

Table 19-1. Example bid schedule

Item no.	Item description	Quantity	Unit	Unit price	Total cost
1	Mobilization/Demobilization	1	Lump Sum	____	____
2	Clearing	1	Lump Sum	____	____
3	Foundation Dewatering	1	Lump Sum	____	____
4	Sediment and Erosion Control	1	Lump Sum	____	____
5	Stripping and Stockpiling Topsoil	4600	Cubic Yard	____	____
6	Demolition	1	Lump Sum	____	____
7	Diversion and Temporary Cofferdam	1	Lump Sum	____	____
8	Unclassified Excavation	10000	Cubic Yard	____	____
9	Borrowing and Placing Embankment Fill	34,000	Cubic Yard	____	____
10	Furnishing and Placing Filter/Drain Sand	1270	Cubic Yard	____	____
11	Borrowing and Placing Road Fill	2600	Cubic Yard	____	____
12	Furnishing and Placing Aggregate Base Course	40	Cubic Yard	____	____
13	Furnishing and Placing Type I Riprap	660	Cubic Yard	____	____
14	Furnishing and Placing Type II Riprap	550	Cubic Yard	____	____
15	Furnishing and Placing Riprap Bedding	350	Cubic Yard	____	____
16	Furnishing and Placing Grouted Riprap	570	Cubic Yard	____	____
17	Outlet Works Modifications	1	Lump Sum	____	____
18	Service Spillway Structure	1	Lump Sum	____	____
19	4-inch PVC Slotted Drain Pipe	600	Linear Feet	____	____
20	Furnishing and Installing 18-inch CMP	50	Linear Feet	____	____
21	Concrete Sill Wall for Emergency Spillway	1	Lump Sum	____	____
22	Instrumentation	1	Lump Sum	____	____
23	Placing and Seeding Topsoil	1700	Cubic Yard	____	____
	Total (Item 1-23)			____	____

description also can indicate what the work scope entails. For example, in Table 19-1, Item 9 "Borrowing and Placing Embankment Fill," it is clear that the embankment fill material will be obtained from an owner-furnished borrow site, and is not an import material. It is also clear that the bid item includes all activities from borrow-pit excavation to final in-place fill after compaction.

Quantity

The quantities indicated on a bid schedule are also referred to as *plan quantities,* which form the basis for the bid. The plan quantities are based on the engineer's best estimate of the work scope and are used for unit-price items. When the actual construction quantities differ significantly from the plan quantities, the contract usually allows the contractor to renegotiate the unit bid price (see Section 22.5). The procedure and criteria for price adjustment for differing quantities are contained in the general conditions and supplemental conditions of a contract and can be different for different owners and agencies. The quantity for lump-sum items, of course, is unity.

Unit

The unit is used for measurement for payment of a unit-price item. Units should be carefully selected and should be practical. For example, most excavation and earthwork are measured with cubic yards, pipes are measured with linear feet, and fabric materials such as geotextile fabric are measured with square yards or square feet.

Unit Price

Unit prices are submitted by a bidder and should include all costs as defined in the payment provisions in the specifications. The costs may include materials, labor, equipment, and the contractor's overhead and profit. The unit price for a lump-sum item is, of course, also the lump-sum price. A definition of what items of work are covered should be included in the payment provisions in the specifications.

Total Cost

The bid price of each item is the product of the plan quantity and the unit price. The final cost is the actual quantity multiplied by the the bid unit prices plus lump-sum items.

During design, it is the design engineer's responsibility to discuss the bid schedule, measurement provisions, and payment provisions with the owner, and the final version of the bid schedule should be agreeable to both parties. Most owners may not be interested in design criteria and details, but they are always interested in how the construction will be paid. Obtaining a concurrence from the owner on the bid schedule during design is particularly advantageous in the event of measurement-related or payment-related disputes with the contractor during construction. The engineer only needs to concentrate on dealing with the contractor and does not need to explain his or her position to the owner.

There are several strategies that can be used to prepare a bid schedule that provides with the best price for the owner and fits within a particular construction budget. Prior to discussing the strategies, the following terms are defined:

Base bid—The base bid is a compilation of contract work items that the owner will be obligated to award to the winning contractor.

Bid alternative—A bid alternative is an alternative work item to replace one or more work items in the base bid. When a bid alternative is used, either the bid alternative or the related base bid item will be awarded, but not both. More than one bid alternative can be used.

Bid additive—A bid additive is an optional work item in addition to the base bid work. At the owner's discretion, a bid additive might be added to the base bid work. More than one bid additive can be used.

A bid alternative is used when the design engineer feels that there is more than one technical solution to a particular feature in the project. The best price or the preferred solution may be determined by the contractors during bidding. By allowing the contractors to bid on an alternative to a base bid design, the designer is soliciting the experience and technical expertise from the contractor to determine what is most cost-effective for the project. Both the base bid work and its alternative should be technically equivalent, allowing an "apple-to-apple" comparison during bid evaluation. Two examples of this strategy are given below.

Dam Drainage Gallery

Several methods (conventional forming, precast concrete panels, and removable "balloon" forms) can be used to construct a drainage gallery in a concrete dam. After the methods are specified, the contractor is invited to bid on any one of the alternatives:

Bid Alternative 8A—Furnishing and Installing Drainage Gallery with Conventional Forms.

Bid Alternative 8B—Furnishing and Installing Drainage Gallery with Precast Concrete Panels.

Bid Alternative 8C—Furnishing and Installing Drainage Gallery with Removable "Balloon" Forms.

Concrete Culverts

New concrete culverts are needed on a small island for drainage improvement, and the island has no facilities for manufacturing concrete. All concrete will need to be shipped a short distance from mainland by boat. Two different methods of culvert design are allowed—cast-in-place concrete, and precast concrete—and the contractor is invited to bid on any one of the alternatives:

Bid Alternative 5A—Furnishing and Installing 4 feet × 6 feet Cast-in-Place Concrete Culverts.

Bid Alternative 5B—Furnishing and Installing 4 feet × 6 feet Precast Concrete Culverts.

With a limited project budget, owners can use bid additives to solicit bid prices on construction work. When favorable bid prices are received for base bid and bid additives that would fit within the project budget, some of the bid additives will be awarded along with the base bid. When high bid prices are received for a base bid and bid additives, only the base bid work will be awarded and the bid additives are not added to the contract. This is the owner's way of "shopping around" for prices, and it is perfectly legal and ethical. An example is given below to illustrate this strategy.

Bituminous Pavement Repair

Two miles of a paved road are in need of repair, but the owner (county) only has funding to repair an estimated 1.5 miles of the road. A base bid is set up for 1.5 miles of road repair, with a bid additive for the remaining 0.5 mile of the road.

Base bid—Paving 1.5 miles of County Road 178 with Asphalt, Sta. 11+50 to 90 + 70.

Bid additive—Paving Additional 0.5 mile of County Road 178 with Asphalt, Sta. 90 + 70 to 117 + 10.

19.3   Methods of Payment

In setting up a bid schedule, there is some work that should be paid on a lump-sum basis, some work that should be paid on a unit-price basis, and some work that can be paid by both methods. A typical civil engineering project contains work items paid using both methods (see Table 19-1). The following are some guidelines on selecting an appropriate payment method for civil engineering projects.

Unit-Price Method

The unit-price method is used when actual quantities are difficult to define during design. This method allows the contractor—without going through change orders—to be paid for the actual quantities encountered during construction. Typical work items paid using this method are excavation, earthwork, foundation grouting, paving, cast-in-place concrete, drain pipes, etc. When a work item is paid using the unit, the paid quantity must be measured for payment. The administration of construction for measurement and payment is beyond the scope of this book.

Lump-Sum Method

The lump-sum method is used when the work performed cannot be measured practically for payment purposes or when the scope of work can be accurately defined. Typical work encountered in civil engineering projects that is difficult to measure for payment purpose includes mobilization, demobilization, foundation dewatering, stream diversion, cofferdam protection, and demolition of existing structures. When the work can be accurately defined, the contractor can prepare an appropriate bid for all of the anticipated costs, and the work item can be performed on a lump-sum basis. For example, if all of the dimensions of a new concrete structure can be identified during design, along with all of the appurtenant components, then this work can be performed using the lump-sum method. An advantage for bidding such a structure as lump-sum work is that there is no need for measurement for payment. Another advantage is that it simplifies the bid schedule.

19.4   Definition of Measurement Methods

When a work item is specified as a unit-price paid item, then the completed work will need to be measured for payment during construction. The measurements can be performed by the contractor or by the owner, depending on the provisions of the contract. It is important to include in the measurement provisions the units and level of accuracy that will be used for measurement. For example, if clearing is being paid according to number of acres, will the area be measured to the nearest acre or to the nearest 0.1 acre? If an earthfill is being paid for by cubic yards, how is the volume measured: in stockpiles, or in place? To avoid disputes during construction, the specifications should state explicitly the method of measurement, as illustrated by the following measurement provisions:

> *Stripping and Stockpiling Topsoil will be measured for payment in cubic yards in the stockpiles after stripping, to the nearest cubic yard.*

> *Embankment Fill will be measured for payment in cubic yards of fill in place at the locations and to the neat line shown on the Drawings, and measured to the nearest cubic yard.*

> *Steel Sheetpile Cutoff Wall will be measured for payment in square feet of wall driven at the locations shown on the Drawings, or as accepted by the Engineer, measured to the nearest square foot.*

> *Backfill Concrete will be measured for payment in cubic yards of concrete placed and approved by the Engineer, and measured to the nearest cubic yard.*

> *Grouting will be measured for payment in cubic feet of grout being pumped into the voids outside of the pipe, to the nearest cubic foot, and approved by the Engineer. Wasted grout not pumped through the grout nipples will not be entitled for payment.*

19.5   Payment of Lump Sum Work

Because the information is important to the contractor for financing the construction and preparing the bid, the method of payment for lump-sum work should be defined in the specifications. Some of these items, such as mobilization, demobilization, and large structures, can be a significant percentage of the project. The contractor is entitled to partial payment of lump-sum items for work completed or for materials and equipment delivered to the site. A typical way to provide progress payment to the contractor for lump-sum items is through milestone completion. When milestone completion is not specified for a lump-sum item, an estimate of the percent completion for that item will be needed for payment purposes. Some examples of schedule of payment provisions for lump-sum items are:

Schedule of payment for Mobilization/Demobilization shall be as follows:

1.   *The total amount of premiums paid by the Contractor to obtain performance and payment bonds will be paid at one time, together with the first progress payment.*

2.   *When 5 percent of the total original contract amount is earned from other schedule items, 50 percent of the amount bid for Mobilization/Demobilization will be paid, less any amount already paid the Contractor for performance and payment bond premiums.*

3.   *When 10 percent of the total original contract amount is earned from other schedule items, an additional 40 percent of the amount bid for Mobilization/Demobilization will be paid.*

4.   *When the Contractor has demobilized from the site, and the project has been inspected and accepted by the Owner, the balance of the amount bid for Mobilization/Demobilization will be paid.*

Schedule of payment for Furnishing and Installing Pumping Wells shall be as follows:

1.   *After the pumping wells outside the drain walls have been installed and tested during Phase 1, 90 percent of the lump-sum bid price will be paid.*

2.   *After groundwater initial drawdown between Phase 1 and Phase 2, 5 percent of the lump-sum bid price will be paid.*

3.   *After all remaining work required for this item has been completed in Phase 2, the remaining 5 percent of the lump-sum bid price will be paid.*

In the two examples above, the payment schedules are somewhat arbitrary but appear to be fair to the contractor. Whether the specified payment schedule reflects the actual costs incurred by the contractor to reach a particular milestone is not as important as defining that schedule in the specifications during bidding. The bidder must properly account for, as finance charges that will be included in the bid, any difference between the amount compensated by the owner and the actual cost. When the payment schedule is clearly defined at the outset, there is no dispute during construction.

19.6   Writing Measurement and Payment Clauses

Measurement and payment clauses for paid items in the bid schedule are typically located in the technical specifications. In the latest CSI specifications format, the price and payment procedures are assigned a Level Two number of 01200 in Division 1. Regardless of where the clauses are located, measurement and payment clauses should accompany every paid item in the bid schedule. The following is a list of guidelines for preparation of these clauses:

- The same item number in the bid schedule should be used in the clauses, and the measurement and payment clauses should be listed in the same order as in the bid schedule.

- A consistent format should be used for each paid item. In particular, two paragraphs should be used: the first paragraph is for measurement, and the second paragraph is for payment. A measurement paragraph is suggested even for lump-sum paid items.

- The measurement provisions should state whether the paid item will be measured for payment. Lump-sum items will not be measured for payment. Unit-price items will be measured for payment, and the unit and the accuracy of the measurement should be specified.

- The payment provisions should list briefly all cost components that will be included under a particular paid item for both unit prices and lump-sum price items, as well as exclusions of work not covered under this item.

The following are some example measurement and payment clauses using the example bid schedule in Table 19-1.

Item 3:   Foundation Dewatering

a.   Dewatering will not be measured for payment.

b.   Payment to the Contractor for dewatering work will be made at the lump-sum price bid in the schedule for Foundation Dewatering. The price shall include all costs for providing materials and labor to install, operate, and maintain all dewatering pumps, including pumps, piping, sump pits and backfill, and other facilities for the control, collection, and disposal of groundwater for the proper construction of all contract work; maintaining foundations and other parts of the work free from water as required; complying with all applicable environmental protection laws and permits, and requirements for operation of the dewatering system; and removing all components of the dewatering system after dewatering is complete.

Item 5:   Stripping and Stockpiling Topsoil

a.   Stripping and stockpiling topsoil will be measured for payment in cubic yards in the stockpile, to the nearest cubic yard.

b.   Payment to the Contractor for stripping and stockpiling topsoil will be at the unit price per cubic yard bid in the schedule for Stripping and Stockpiling Topsoil, which price shall include all costs for equipment, materials, and labor to mow grass and small shrubs, strip, and stockpile topsoil.

Item 8:   Unclassified Excavation

a.   Unclassified excavation will be measured for payment in cubic yards, to the nearest cubic yard, of materials excavated in place to the limits of excavation shown on the Drawings, or otherwise approved in writing by the Engineer.

b.   Payment to the Contractor for authorized excavation will be at the unit price per cubic yard bid in the schedule for Unclassified Excavation, which price shall also include stockpiling of materials to be reused, and disposal of excavated materials not suitable or required for construction. Unauthorized excavation will not be paid for.

Item 17:   Outlet Works Modifications

a.   *Outlet works modifications will not be measured for payment.*

b.   *Payment to the Contractor for outlet works modifications will be at the lump-sum price bid in the schedule for Outlet Works Modifications, which price shall include all labor, materials, and equipment to place the upstream concrete collar; grouting of the existing 36-inch pipe; new 24-inch PVC pipe; concrete pipe encasements and joints; concrete outlet basin; and all concrete testing.*

Item 21:   Concrete Sill Wall for Emergency Spillway

a.   *Concrete sill wall for emergency spillway will not be measured for payment.*

b.   *Payment to the Contractor for concrete sill wall will be at the lump-sum price bid in the schedule for Concrete Sill Wall for Emergency Spillway, which price shall include all labor, materials, and equipment to furnish and construct cast-in-place concrete wall, and concrete testing. Excavation to construct the wall will be separately paid for.*

CHAPTER 20

PRESENTING REFERENCE DATA

20.1   General

Rarely are heavy civil construction projects constructed based on design drawings and technical specifications only. In most cases, data collected during the investigation and design phases are furnished to the contractor for his or her use. These data include topographic survey, underground and existing structures and utilities, subsurface data, geologic data, laboratory and field test data, precipitation and weather data, and stream flow data. The importance of these data in civil engineering projects is discussed in Chapter 3. This chapter discusses the presentation of these reference data in construction documents.

It is important to point out that, when data are included in the construction documents, these data become part of the contract, and as such, should be factual, accurate, and obtained with the same care and competency with which drawings and technical specifications are prepared. Any errors or omissions on these data are grounds for changed condition claims during construction.

Most of these data, such as subsurface data and test data, are obtained as part of the engineering investigations required for design. These data are typically summarized and presented in engineering reports. These reports also contain conclusions of findings, recommendations for design, considerations for construction, etc., that represent interpretations, judgments, and evaluations by the engineer. Some owners and design engineers include these reports in their entirety in construction documents, with qualifications that the contractor should provide his or her own interpretations and evaluations on the data provided. This practice assumes that the contractor can separate factual data from the engineer's interpretations and evaluations. When the separation between data and interpretations is not straightforward and explicit, it allows the possibility of the contractor to incorporate the engineer's interpretations and evaluations in planning his or her own work or preparing his or her designs (e.g., in performance specifications).

Construction documents furnished to the contractor should only contain factual data, not the engineer's interpretations, evaluations, and recommendations to the owner. When engineering reports contain site investigation data and test data that are relevant and important to the performance of the construction work, those data should be separated from the report before amending the construction documents. This chapter contains suggestions on presenting data in construction documents for civil engineering projects.

20.2   Relevant Data

The following is a list of data that are typically collected for civil engineering projects. This list is not intended to be exhaustive, as each project may have other unique and special data requirements not included on this list.

Subsurface Data

Subsurface data include drill logs from boreholes, logs from test-pit excavations, field test data (standard penetration tests, field permeability tests, etc.), groundwater conditions in boreholes and test pits, observation-well installation and monitoring data, laboratory test data, and geophysical field investigation data. Some of these data can be conveniently included in construction drawings (see Section 5.3) by using simplified graphic logs. In most cases, subsurface data cannot be conveniently summarized graphically, and should then be included as part of the technical specifications. Data that are commonly included in specifications include detailed field logs, field permeability values, water levels measured at the time of drilling (with appropriate disclaimer clause), observation-well installation reports, and laboratory test reports.

Test Data

Results of field and laboratory test data collected during field investigations and laboratory testing are important information that the contractor needs for his or her work. For example, when an owner specifies that an earthfill borrow area should be used for a project, the engineer should furnish the relevant soil properties (natural water contents, plasticity, compaction characteristics, etc.) to the contractor. Field permeability test results are useful to the contractor for planning his or her unwatering and dewatering operations during foundation excavation. Field anchor test results for a clay shale foundation are useful to the contractor for designing the foundation anchors that are typically specified using performance-based requirements. Soil-cement mix design results are useful to the contractor for planning his or her material handling and mixing operations. When an owner-furnished quarry will be used as a source of riprap rock, the results of a test-blast program will be useful to the contractor for planning his or her quarry blasting operations during production of the riprap.

Hydrologic/Climatic Data

All heavy civil construction takes place outdoors, and planning construction work to account for climatic factors (precipitation, freezing conditions, storms, etc.) is just one of the risky elements inherent in this type of construction. In some cases, when work is at or near a stream, the risk of flooding at certain times of the year should be accounted for by the contractor. It is the owner's responsibility to gather as much of the climatological data as possible before construction, and to provide these data to the contractor. However, some owners and engineers do not agree with this approach, and would leave this type of information entirely to the contractor, thus requiring the contractor to absorb more risk.

Meteorological data can be obtained from the National Climatic Data Center of the National Oceanic and Atmospheric Administration. Information typically provided from this source includes mean maximum and minimum monthly temperatures, degree days, and precipitation records for the reporting station closest to the project site. Stream flow records are available from the U.S. Geological Survey for the gaging station closest to the project site. It is common to show stream flow records in graphical form, which is then inserted into the construction drawings set. It is preferable to present the data in table form in the specifications rather than in graphical form in the drawings to avoid any plotting errors introduced by the engineer. It is unnecessary to present both sets of duplicate information.

Records of Existing Structures

Many civil engineering projects involve rehabilitating or upgrading existing structures. Any available records on these existing structures furnished by the owner to the engineer for design should also be furnished to the contractor. These records are important to the design engineer for technical and constructability reasons, but they are also equally important to the contractor. If the structures are to be demolished, that information would provide the contractor with the level of effort and equipment needed. If the structures are to be modified, that information would help the contractor understand how the new work would fit into the old structure, especially when some of the work would most likely be field-fit. Equally important to the owner and contractor is how the information is used to protect the existing structure that remains.

Frequently, records of existing structures include record drawings. Record drawings can be included in the construction drawings set to supplement data that are referenced in the specifications.

It is the responsibility of the contractor to review all of the reference data furnished by the owner. A contractor who fails to do so would lose his or her claim rights. Legally, contractors are held to the information and data in the specifications contained in the contract, regardless of whether they read them. This issue had been challenged in court, and has been ruled in favor of the owner by the U.S. Court of Appeals (American Society of Civil Engineers, 2002). Therefore, contractors should be aware of the consequences of neglecting all of the relevant data in the contract.

20.3   CSI Format

It is unclear where the relevant data listed in Section 20.2 can be presented under the current format of the CSI format (see Chapter 18). In the absence of specific guidelines from the CSI, a new division in the specifications has been used, designated as *Division 17: Data*, to present reference data in technical specifications. The following section numbers are proposed for this division:

17100: Subsurface Data

17200: Field and Laboratory Data

17300: Hydrologic Data

17400: Climatological Data

17500: Data of Existing Facilities

As seen in the list above, there is a significant amount of flexibility for a designer to work within this new division. It remains to be seen how CSI will address the organization of this type of information in technical specifications.

20.4 Examples

The Appendix contains several example sections containing relevant data for a typical heavy civil construction project. For simplicity, only the three-part write-up for each section is shown in the examples, and the data themselves are not included. By nature of this type of section, only *Part One: General* is applicable, and the remaining two parts are usually not used, as shown in the examples.

PART 4

COST ESTIMATE

CHAPTER 21

PURPOSE AND USE

21.1   Introduction

In a broad sense, an engineer's construction cost estimate is a designer's prediction regarding the probable cost of a construction project. Traditionally, it has always been part of an engineer's responsibility to assist the owner to plan and budget the money (cash flow schedule) needed for a construction project. Based on the engineer's estimate, the owner obtains the money necessary for construction. It is obvious that the engineer's cost estimate should be confidential prior to bid opening, and the information should not be shared with anyone interested in bidding on the construction of that project. Depending on a variety of factors that will be discussed in the next few chapters, the actual construction cost may or may not be close to the engineer's construction cost estimate. Disputes or claims related to accuracy of engineers' estimates, by project owners against designers, occur from time to time. Recently, liability-conscious design professionals have stopped using the term "engineer's cost estimate," and replaced it with the term "opinion of probable construction cost" (Dixon 1998). Changing the way we label estimated construction cost does not make the problems go away or decrease the engineer's liability. What is important is for an engineer to perform this work in a professional and competent manner, similar to the work he or she has done in preparing the construction drawings and technical specifications. While the estimated cost of a construction project is really an engineer's professional opinion, there is nothing inherently wrong with the term "engineer's cost estimate" if the assumptions, guesses, and uncertainties are adequately discussed with the owner. In this book, the traditional term will be used for this process.

There are two separate steps in an engineer's cost estimate. The first step is estimating the quantities based on the design. Because they are calculated based on the layouts on the drawings (or plans), these quantities are also referred to as *plan quantities*. The second step is estimating the prices associated with estimating quantities of construction work items. The two sets of data are combined to obtain the total cost of the construction.

21.2   Levels of Estimate

Just as there are levels of design, there are several levels of an engineer's cost estimate (see Section 4.2). Because different contingency factors and unknowns are associated with each level, it is important to distinguish between the levels of a cost estimate. In the early phase of a project, an engineer's cost estimate is performed to compare costs of various technically feasible alternatives; this level of cost estimate is called a *planning-level cost estimate* or *feasibility-level cost estimate*. After a preferred alternative has been selected, an in-depth, conceptual design is performed to obtain a *conceptual-level cost estimate* or *preliminary-level cost estimate*. Frequently, the owner uses the cost information and contingency factors recommended by the engineer at the conceptual level to obtain funding for the final design and construction.

The importance of a conceptual-level cost estimate cannot be over-emphasized. On one hand, the engineer only has limited design data and information to arrive at the funding-level cost estimate. It is at this level of cost estimate that the experience and judgment of the engineer play a significant role in the successful planning of the project. Simply because the design is conceptual or preliminary, and nothing will be constructed from this design, many owners and design professionals underestimate the importance of using experienced designers for designing and estimating the cost of the project at an early phase. Inexperienced designers can commit significant errors, and large-cost items can be neglected in the cost estimate. Sometimes, these errors are not realized until final design, and sometimes they are not apparent until construction. The consequences of underfunding a construction are more serious in federal construction projects than in private-sector projects. When underfunding for a federal project is discovered during final design, the project cannot go into bidding and construction. In such a situation, additional funding needs to be reallocated under the current budget, or it needs to be approved by Congress during the annual budgeting process, which results in delays. Sometimes, designs must be revised to downsize the project in accordance with available funding. Neither prospect is attractive to federal agencies, and sometimes underfunded projects may simply be terminated.

When a concept goes into final design, plans and specifications are prepared, and a final design level engineer's cost estimate is prepared. Most projects go directly into bidding, selection of a contractor, and construction. Other than turnkey projects, the engineer's cost estimate at this level serves many purposes:

- The engineer's cost estimate at this level is compared with the funding-level cost to evaluate whether adequate funding is available to complete the construction.

- The plan quantities on the bid schedule are used as a basis for bidding.

- The prices estimated by the engineer are used to evaluate the contractor's bids.

21.3   Role and Responsibility

Estimating the cost of a construction project can be handled in three ways:

1. The engineer performs all of the cost-estimating work.

2. A professional cost estimator performs all of the cost-estimating work.

3. A team of the engineer and professional cost estimator, depending on the level of development, performs the cost-estimating work.

The advantages and disadvantages of each arrangement are discussed below.

By Engineer Only

This arrangement is common in small- to medium-sized firms that do not have their own in-house cost estimators. Typically, staff engineers will calculate quantities, and more senior engineers will estimate the prices (see Chapter 23). Historically, this approach has resulted in cost estimates that may vary widely compared with bids and actual construction costs. Most engineers are not trained to estimate construction costs. For example, even experienced design engineers are not knowledgeable in the production rates of some construction crews and equipment.

When an engineer is uncertain about costs, the natural tendency is to be conservative. Although conservatism is essential for technical design and analysis, an overly conservative cost estimate is not always advantageous for the owner. An overly conservative cost estimate may have the following consequences:

- The owner's financial restrictions may simply terminate the project.

- The fact that all of the bids are significantly lower than the engineer's cost estimate may raise some questions regarding the competency and credibility of the estimating engineer.

- Over-financing of a project may cause financial problems for the owner.

- For state and federal projects, it may create problems allocating and managing all the money left over from overfunding.

By Professional Estimator Only

Some large design firms are staffed with in-house, professional cost estimators for the specific purpose of performing construction cost estimating. Some of these specialists are former contractors and may or may not have any technical training and background. Two scenarios are possible regarding estimation of cost of a design project:

- The estimator is handed a set of plans, specifications, and a bid schedule, and he or she is required to perform all of the quantity takeoffs and estimate the prices of the items in the bid schedule.

- The estimator is handed a set of plans, specifications, a bid schedule, and plan quantities, and he or she is required to estimate the prices in the bid schedule.

Regardless of whether the estimator is given the quantities, he or she may approach the estimate in a similar manner as to the bidders. The main difference is that he or she is not bidding on the project. His or her goal is to target within the lower to middle range of the anticipated bid price. In any case, the estimator usually is not involved during the design and, therefore, does not know the design intent, design criteria, site conditions, or site constraints. However, he or she must work with the design engineer to understand clearly the known and unknown design-related conditions to give adequate unit prices or lump-sum prices.

The use of professional estimators in cost estimating represents a significant improvement in the accuracy of the engineer's cost estimate, but that capability is reserved for a limited number of large firms with in-house estimators. The lack of a site visit by the estimators is a disadvantage, and the designers may not give the estimator all of the necessary information to estimate effectively the cost of construction.

Team of Engineer and Professional Estimator

To utilize the expertise and knowledge of a professional cost estimator, an engineer should involve the estimator during design, and then provide him or her with all of the pertinent information for the cost estimate. Specifically, the engineer should:

- Allow the cost estimator to visit the project site. If bidders are given an opportunity to visit the site in a prebid meeting, why should the professional cost estimator be denied that opportunity?

- Provide pertinent information to the estimator—designer's assumptions on construction methods, sources of materials, preferred manufacturers and suppliers, owner's requirements (such as requirement of using union labor), schedule of construction, etc. Note that this information is privileged to the estimator only, but not to the contractors bidding on the project.

When engineers estimate the cost of a project, they can seek a second opinion by engaging an outside professional construction cost estimator. This outside consultant essentially acts as a reviewer for the engineer and, in most cases, would make some contributions toward the quality of the plans, specifications, and constructability in addition to cost information.

CHAPTER **22**

QUANTITY ESTIMATE

22.1   Units

In civil engineering design and construction, quantities are estimated in units of length, distance, area, volume, or weight. The selection of units of measurement is based primarily on ability to measure units practically and quickly during construction for payment purpose. When it is difficult to quantify the work with measurable units, as with foundation dewatering, it is customary to use a lump-sum method for payment (see Section 19.3).

Measurement of Length

Using length to represent work effort is suitable when the cross section is relatively uniform along the limits of the work. The most common unit of length measurement is linear feet (LF). Examples of construction work that can be represented in linear feet are:

Length of pipes (drainpipe, sewers, culverts, storm drains)

Length of drilling (drain holes, grout holes)

Length of fence, guardrails, or handrails

Length of wall with uniform section

Length of tunnel

Pilings

Measurement of Area

Using area to represent work effort is suitable when the thickness or depth is uniform within the limits of the work. The common units of area measurements are square feet (SF), square yards (SY), and acres (AC). The use of square feet or square yards is a designer's preference, as the two units only differ by a factor of nine. When the anticipated quantity is large, say in the tens of thousands of square feet, it is more practical to use acres or thousand-square-foot

(MSF) as the unit of measurement. Examples of construction work that can be represented in these units are:

Area (SF or SY) of fabric (geotextiles, membrane, erosion-protection mat)

Precast concrete panels (SF or SY)

Shotcrete or gunite (SF or SY)

Steel sheetpiles (SF)

Proof-rolling subgrade (SF)

Pavement (SF)

Clearing and grubbing (AC or MSF)

Site grading (AC or MSF)

Reclamation of disturbed areas (AC or MSF)

Measurement of Volume

When the geometry of work is such that length and area are not suitable to represent the work effort, such as in excavation and backfill, then it is best to use volume as a unit of measurement. Most volumes are measured in place. For example, excavation is typically measured as the difference between the existing ground surface before excavation and the limits of final excavation. Backfill is typically measured in place after compaction. When approval of materials in stockpiles is necessary, it is also customary to measure the materials in the stockpiles. To avoid disputes and claims during construction, it is important to specify in the measurement and payment clauses the method of measurement (see Chapter 19).

The most common unit of volume in civil engineering construction is the cubic yard (CY). Smaller units, such as the cubic foot (CF), are used for concrete materials such as grout. Examples of volume measurement are:

Excavation, earth fill, riprap (CY)

Conventional formed and unformed concrete (CY)

Grout (CF)

Measurement of Weight

Theoretically, volume and weight can be used interchangeably if the unit weight is known. In cases in which material or product is transported by trucks, the weight is more readily available because the trucks are weighed. The most common unit of weight is the ton (2000 pounds). Contractors typically purchase fabricated metals from fabricators in pounds. However, because the fabricated products usually are not weighed individually after fabrication, the owner typically pays for the metals as lump-sum items. When reinforcing steel is purchased separately, however, it is usually paid for in pounds. Riprap is also measured and paid for in tons; in fact, contractors prefer to be paid this way because it eliminates the need for conversion from weight (easily obtained from truck scales in quarries) to bulked cubic yards in place. The issue of weight or volume for riprap is one of designer preference.

Examples of weight measurement are:

Riprap (tons)

Asphalt (tons)

Bulk cement and fly ash (tons)

Cement for grouting (94-pound bags)

Processed concrete aggregate (tons)

Reinforcing steel (pounds)

Other Units

Occasionally, construction work effort cannot be represented easily by length, area, volume, or weight. Examples of other units include:

Per each (e.g., survey monuments, observation wells, anchors, hook-ups for grouting)

Hour (e.g., water-pressure testing, standby time, crew and equipment time)

Day (e.g., pumping)

22.2   Quantity Calculations

Quantity calculation (also called quantity takeoff) is part of the engineering design process and, as such, should be performed with the same care and manner as other design documentation. It is suggested that the following general procedures be followed:

- The level of design, drawing revision number, bid item number, and date of calculation should be clearly stated on all calculation sheets.

- The method of measurement and calculation should be stated. When a planimeter is used, the data obtained from the planimeter should be recorded. When CAD software is used, computer printouts should be attached and properly labeled.

- The same terminology used in the specifications and drawings should be used in the calculations. For example, if there are three different types of backfill (e.g., select fill, general fill, structural fill), each type of fill should be clearly differentiated and identified.

- The source of dimensions should be stated clearly. When dimensions are obtained directly from design drawings, the sheet number, cross-section designation, or stations should be referenced. When dimensions are assumed because they are not available on the drawings, the assumptions should be stated. This is particularly helpful for someone who is checking and reviewing the calculations. In fact, the person performing the calculations should always ask the question: *Have I provided enough information for the checker?*

- When additional drawings or sketches are required to improve accuracy, the supporting drawings should be attached.

- Cost allowance should be made to account for unavoidable waste and overlap during construction. For example, when geotextile fabric is installed, the panels are overlapped by an amount recommended by the manufacturer. The actual fabric area used is always larger than the finished area shown on the drawings. This cost allowance is needed even if the fabric is measured as neat area without overlap.

- The estimator should refrain from excessive rounding of calculated quantities. In general, rounding should be performed at the end of calculation for a particular item, and not during intermediate steps. For example, it is appropriate to round off a number from 7908 CY to 7900 or 8000 CY, but not from 7122 CY to 8000 CY. Excessive rounding will increase the chance for a quantity to exceed the allowance during construction (see Section 22.5).

- Ample space should be reserved in the calculation sheets for the checker's comments and additional calculations if necessary.

- As he or she would when checking other engineering calculations, the checker should use a pen of a different color to distinguish the original calculation from the checker's comments. The checker also should initial and date the checked calculations for tracking.

Quantity calculations are required for all unit-price items on the bid schedule and may be required to support some lump-sum items. Examples of lump-sum items that require quantity takeoff to support the price estimate include:

- Volume of concrete for a reinforced concrete structure

- Weight of fabricated metals

- Breakdown of components for dewatering systems, stream diversion, cofferdam protection, etc.

- Breakdown of components for reclamation features—regrading area, topsoil quantity, seeding and mulching area, pavement repair, etc.

22.3   Methods of Computations

It is assumed that the reader is knowledgeable of basic mathematical skills and formulas to compute areas and volumes of commonly encountered shapes and geometric configurations. Mensuration formulas, readily available in mathematics textbooks and other engineering handbooks, are not presented in this book. This section addresses the methodology of using information from design drawings in quantity computations and the tools used to complete such computations. In particular, this section discusses the method of estimating quantities of irregularly shaped geometrical design features, for which rigorous mathematical treatment is inappropriate or impossible.

Length or Distance

Length or distance can be measured from the drawings with an engineer scale, or a map-measuring device. On design drawings, linear features—pipes, culverts, roads, tunnels, etc.—are shown on plan views with control survey information, such as stations, curve lengths, northings, and eastings. When this type of information is shown on the drawings, it is recommended that the survey control data or dimension callouts are used directly in the calculation of the quantities, rather than measuring the features. When depths of drill holes (for drainage or grouting) are shown on the drawings, the depths called out on the drawings should be used instead of direct measurement. Direct measurements are required when drawings are still in their feasibility or conceptual-design stage or in their early final design phase, when controls have not been established.

Area Calculation

When an area to be calculated can be subdivided into regular shapes (triangles, trapezoids, etc.), the area is calculated using appropriate mensuration formulas. Again, dimensions directly called out on the drawings should be used and supplemented with other distances that can be measured with a scale. Generally, however, areas are irregularly shaped in civil design, and a planimeter can be used. A planimeter is a mechanical device that can be used to determine the area of a figure by moving its tracing point around the perimeter of the area. Although the accuracy of this instrument is independent of the irregularity of the area to be estimated, there is some difference in the result from different operators or even from different trials of the same operator. Therefore, it is recommended that each area should be traced at least twice, and the average of the measurements should be used to determine the area. Modern planimeters have digital displays and are capable of measuring areas in which the vertical scale is different from the horizontal scale. Figure 22-1 is an illustration of area determination using a digital planimeter.

When a drawing is drawn in CAD, areas can be determined with extreme precision using a computer. As discussed in Section 22.2, all computer-determined quantities should be documented with a computer printout and checked by hand by the engineer. Checking by hand

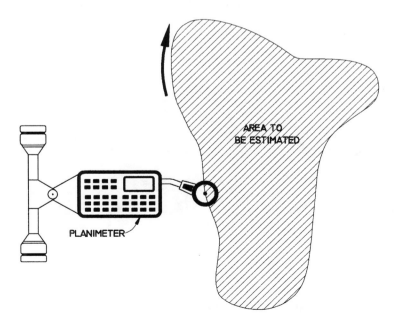

Figure 22-1.   Measuring area with a planimeter.

ensures that the CAD drafter input proper area boundaries and scales. This check can be very coarse and is only for confirming the general value of the CAD printout data (±5% accuracy).

Volume Calculation

When a volume to be calculated can be subdivided into regular shapes (pyramids, cubes, or other prismatic components), the volume is calculated using appropriate mensuration formulas. Again, dimensions directly called out on the drawings should be used and supplemented with other distances that can be measured with a scale. Generally, however, areas for which volumes must be determined in civil design are irregularly shaped and other approximate methods are used. Methods to determine volume include the subdivision method, prismoidal method, average end-area method, and contour method. A comprehensive treatment of these methods can be found in Church's *Excavation Handbook* (1981).

By far, civil engineers most commonly use the average end-area method to determine volumes for quantity estimate. This method is popular because of its simplicity and acceptable precision for long structures (e.g., roads, levees, pipelines). With this method, the volume of a three-dimensional object with parallel or near-parallel cross sections at each end is equal to the average of the cross-sectional areas at the two ends multiplied by the distance between the sections. Figure 22-2 illustrates this method. Cross sections at some regular intervals, such as 50 feet or 100 feet, can be selected and drawn, and the areas of these cross sections are measured, most commonly with a planimeter (see "Measurement of Area" in Section 22.1). Some useful guidelines for this method are given below:

- The number of cross sections required depends on the changes in cross sections of a particular feature. When the cross sections are relatively uniform, fewer cross sections are required. More cross sections are required to characterize the change in cross sections. Most design drawings only contain a typical cross section that can be used to generate additional sections for quantity estimate. These additional sections do not need to be included in the design and construction drawings, but should be attached to the computation sheets for checking and future reference.

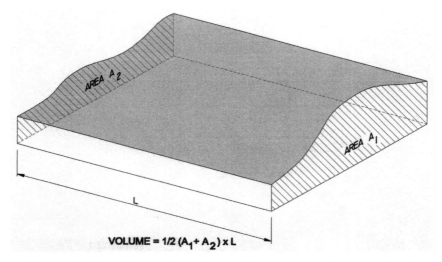

Figure 22-2. Measuring volume with average end-area method.

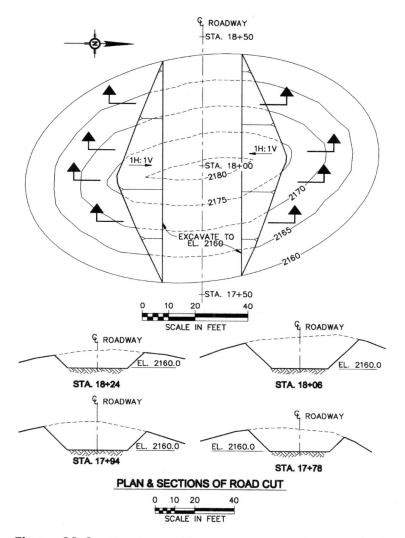

Figure 22-3. Road cut to illustrate average end-area method.

- The accuracy of this method depends on the distance between the cross sections. The smaller this distance, the more precise will be the volume estimate, but—because more sections are involved—the effort will be more time-consuming. Therefore, the estimator will need to decide on a balance between precision required and the time and budget available for this work.

As an example of the average end-area method, Figure 22-3 shows a plan and cross sections for a road cut through a knob in bedrock. Four cross sections are drawn, and the cross-sectional areas of the excavation are measured with a planimeter. The measured sectional areas and the calculations are shown in Table 22-1. The estimated excavation quantity for the road cut in this example is 1277 cubic yards, which can be rounded off to 1280 cubic yards.

Table 22-1. Illustration of average end-area calculation for road cut excavation

Station	Areas (sq. ft)		Distance between sections (ft)	Volume (cubic yards)
	Section area	**Average area**		
17+60	0			
		203	18	135.3
17+78	406			
		581	16	344.3
17+94	756			
		810.5	12	360.2
18+06	865			
		562.5	18	375
18+24	260			
		130	13	62.6
18+37	0			
Total Volume (cubic yards)				1277.4

22.4 Earthwork Calculations

Because the densities of earthwork materials will change during excavation, handling, and placement, calculations related to earthwork—excavation, fill, riprap, etc.—differ from quantity estimates for other materials (e.g., concrete and steel). Regardless of the method used to compute the quantities, the following changes in volumes should be accounted for in earthwork calculations:

- Materials excavated from natural sources (*bank-run* or *pit-run materials*) will increase in volume in the haul trucks and/or in the stockpiles. This increase in volume is known as *swell*.

- Materials that are compacted by mechanical equipment will decrease in volume when compared with loose stockpiles and sometimes when compared with bank-run materials. This decrease in volume is known as *shrink*.

Another problem unique to earthwork relates to how it is paid for during construction. When aggregates, filter materials, or riprap are paid for by weight (see Section 22.1), conversion factors must be used to convert the in-place quantities into weights. Regardless of how these materials are paid, either the engineer or contractor needs to use conversion factors for the following reasons:

- When materials are paid for by weight, the engineer needs to convert in-place volumes to weights. Because these materials are purchased and hauled based on weight, the contractor does not need the conversion process. The contractor prefers this method.

- When materials are paid for by in-place volume, the engineer does not need to use the conversion factors; that responsibility is shifted to the contractor. Not only does the contractor have to convert the materials from weights to volumes, he or she also must estimate the effects of swelling and shrinking during borrowing, stockpiling, and compaction and losses during the handling process. The engineer prefers this method.

Swell factors and shrink factors can be used to account for changes in densities of excavated and compacted earthwork. Civil and mining engineers have used these factors for more than 100 years. A complete list of swell and shrink factors for a variety of materials, including soils, bedrock, ore materials, peat, and caliche, can be found in Church's *Excavation Handbook* (1981). Only factors related to soil materials, aggregates, and riprap are presented in this book.

Published swell and shrink factors are typically defined as follows:

$$\% \text{ Swell} = \frac{(\text{bank run density} - \text{density in loosened conditions}) \times 100}{\text{density in loosened conditions}} \quad (22\text{-}1)$$

$$\% \text{ Shrink} = \frac{(\text{bank run density} - \text{compacted density}) \times 100}{\text{compacted density}} \quad (22\text{-}2)$$

As defined above, the percent swell is generally a positive number, because most bank run densities are higher than the densities in loosened or stockpile conditions. For soil materials, including clays, silts, sands, and gravels, the percent shrink is generally a negative number, because the compacted density is generally higher than bank-run density.

Current published tables of swell and shrink factors do not include conversion from loosened conditions (e.g., in stockpiles or in loaded trucks) to compacted condition. A third conversion factor, *compact*, is proposed for this scenario. The compact factor is defined as follows:

$$\% \text{ Compact} = \frac{(\text{density in loosened conditions} - \text{compacted density}) \times 100}{\text{compacted density}} \quad (22\text{-}3)$$

As defined in this way, the percent compact can be calculated easily from the percent swell and shrink factors in published tables as follows:

$$\% \text{ Compact} = \frac{(\% \text{ Shrink} - \% \text{ Swell}) \times 100}{\% \text{ Swell} + 100} \quad (22\text{-}4)$$

For example, if the percent shrink is –10%, the percent swell is 35%, the corresponding percent compact is:

$$\% \text{ Compact} = \frac{(-10 - 35) \times 100}{35 + 100}$$
$$= -33\% \quad (22\text{-}5)$$

Table 22-2 contains the percent swell, shrink, and compact factors for commonly encountered soil materials.

The factors in Table 22-2 should be used with caution for the following reasons:

- By virtue of the different natural methods of formation, deposition, and soil composition, bank-run soil materials are highly variable. The percent swell and percent shrink shown in the table are dependent on the bank-run conditions and represent only the average conditions.

- Loosened conditions are also highly variable, depending on a number of factors, such as method of excavation, moisture condition of the soil, and cohesion characteristics. The percent swell and percent compact in the table, which are dependent on the loosened conditions, represent only the average conditions.

- Compacted density is also highly variable, depending on a number of factors, such as the loose-lift thickness, size and type of the compaction equipment, and in-place moisture content. The percent shrink and percent compact shown in the table are dependent on the compacted density and represent only the average conditions.

In general, the factors in Table 22-2 are within the level of accuracy typically expected for an engineer's cost estimate. However, to better characterize the changes in densities of the various earthwork materials, a contractor planning his or her borrowing, stockpiling, hauling, and compaction operations—especially for large projects—should perform project-specific field and laboratory tests, such as test fills using proposed equipment and in-situ density measurements of bank-run materials.

22.5   Allowance for Quantity Difference

This section discusses the implications of the difference between estimated quantity and the actual quantity required during construction. There are many reasons that actual quantity of a particular work item can differ from the estimated quantity, such as:

- Accuracy of the quantity estimating method—The accuracy of different methods of quantity calculation will vary. For example, areas and volumes determined with the planimeter method typically will have an accuracy of $\pm 0.5\%$ to $\pm 1.0\%$. Areas and volumes determined using CAD software will yield a better accuracy, but the results are only as good as the input data.

- Errors in quantity calculations—Errors can be committed by engineers and CAD drafters performing the calculations in many ways—inaccurate measurements, arithmetic errors, wrong equations, wrong scale, etc. As suggested in Section 22.2, all quantity calculations, hand calculations, and computer calculations should be checked to avoid errors. It is wishful thinking to assume that computer-generated calculations will always be accurate and do not require checking.

- Field conditions that differ from those assumed in design—Field conditions that can result in different quantities of earthwork include different existing topography and different subsurface conditions that require different depths of excavation. This source of difference can be minimized by adequate site characterization effort during investigations and design. Some site work quantities, such as excavation, are based on the designer's assumptions regarding behavior of materials during construction. If the designer assumes that the ground can be excavated in a 1H:1V

Table 22-2. Earthwork quantity conversion factors

Material	Unified soil classification	Volume change factors			Weight-volume conversion (ton/cu.yd)	
		%Swell	%Shrink	%Compact	Loose	Compacted
Gravel, Clean, Moist, Poorly-Graded	GP, GP-GM, GP-GC	9	−7	−15	1.5	1.8
Gravel, Clean, Moist, Well-Graded	GW, GW-GM, GW-GC	9	−11	−18	1.6	1.9
Gravel, Dirty, Moist, Silty Fines	GM	9	−11	−18	1.5	1.8
Gravel, Dirty, Moist, Clayey Fines	GC	14	−9	−20	1.4	1.8
Sand, Clean, Moist, Poorly-Graded	SP, SP-SM, SP-SC	10	−15	−23	1.4	1.8
Sand, Clean, Moist, Well-Graded	SW, SW-SM, SW-SC	9	−18	−25	1.4	1.9
Sand, Dirty, Moist, Silty Fines	SM	5	−16	−20	1.4	1.7
Sand, Dirty, Moist, Clayey Fines	SC	5	−15	−17	1.4	1.7
Silt, Moist, Low Plasticity	ML	11	−9	−18	1.2	1.5
Silt, Moist, High Plasticity	MH	12	−10	−20	1.1	1.4
Clay, Moist, Low Plasticity	CL	33	−8	−31	1.2	1.8
Clay, Moist, High Plasticity	CH	33	0	−25	1.2	1.6
Aggregate Base Course	GM, SM					
Dry		17	−18	−28	1.5	1.9
Moist		14	−16	−27	1.5	2
Filter/Drain Sand	SP, SW					
Dry		17	−21	−31	1.2	1.8
Moist		16	−19	−30	1.3	1.8
Topsoil	OL	50	−22	−48	0.8	1.6
Riprap						
D50 = 6 inches						1.45
D50 = 9 inches						1.49
D50 = 12 inches						1.5
D50 = 18 inches						1.89
D50 = 24 inches						2.2

slope, but the actual ground is so loose and sandy that a 1.5H:1V slope is the steepest possible, then the quantity of excavation will be underestimated, as will the required backfill.

- Design modifications during construction—When design modifications are made during construction, quantities are usually affected by the change. Design changes during construction are handled through change orders.

Most construction contracts contain a provision to handle changes in quantities for unit-price work items. When the actual quantity of a particular bid item is within a certain percentage of the original bid quantity (or plan quantity), there is no change in the unit price for that item. When the actual quantity is less than or more than the allowable percentage, the contractor is entitled to renegotiate the unit price for the item. There is no fixed allowable percentage, and that number varies from contract to contract, and from owner to owner. The EJCDC General Conditions defer that option to the Supplemental Conditions, and the Supplemental Conditions allow the owner to insert whatever percentage is appropriate. Typically, the allowable percentage ranges from 15% to 30%. Because of the inherent unknowns and risk, an allowable percentage of 25% is recommended for civil engineering construction.

When the actual quantity is significantly less than or more than the plan quantity, the contractor's cost to perform the work is expected to change. Generally, when the actual quantity is significantly less than the plan quantity, the actual unit cost to the contractor increases, and the contractor is therefore entitled to renegotiate for a higher unit price than the bid price. When the actual quantity is significantly more than the plan quantity, the actual unit cost to the contractor decreases and the owner is therefore entitled to renegotiate for a lower unit price than the bid price. However, large increases in quantities can have an impact on construction schedules, and the contractor is also entitled to a schedule extension. Although there is a contractual system to handle changes in quantities and unit price, renegotiation for unit prices should be kept to a minimum during construction. To minimize significant changes during construction, it is the responsibility of the design engineer to estimate the quantities as accurately as possible based on the design and site information before the start of construction.

22.6 Quantity Survey

When construction work is paid for by unit price, the actual quantities should be measured during construction for payment purposes. The limits of construction work for which the contractor is entitled to payment are shown on the drawings as *neat lines*. It is important to show neat lines in excavations and backfill. For example, when the contractor excavates beyond the neat line, the overexcavation is not entitled to payment, unless the engineer authorized it in advance or the engineer directs the overexcavation in the field because of unanticipated conditions, such as weak soils. The neat lines for concrete structures coincide with the dimensional limits of the structure.

A designer should be careful in calling out minimum dimensions, such as "5 feet minimum excavated width." The contractor excavates 10 feet, which meets the design requirement, and demands payment for 10 feet of excavation. If five feet is the intended excavated width, then that exact dimension should be called out to avoid paying the contractor for excessive quantity.

Typically, the contractor is responsible for surveying or measuring the completed work, and sometimes the owner will also engage his or her own surveyor to check the contractor's quantity data. Quantity survey is time-consuming and disruptive to construction. Excessive effort to measure quantities for progress payment is also costly, regardless of who does it. Therefore, it is important during design to carefully select the units in the bid schedule and measurement and payment clauses to make quantity survey as simple and as quick as possible. Some examples illustrating this principle are given below:

- Stripping of topsoil is required, but the thickness of the topsoil is highly variable. Instead of measuring the actual thickness of the topsoil that is stripped, it is more practical to measure the topsoil in the stockpile for payment purposes.

- Reinforced concrete contains rebars, joint materials, waterstops, etc. Instead of measuring the volume of the concrete, weight of rebars, and lengths of the joint materials and waterstops and paying for them separately, it is quicker to measure the in-place quantity of the concrete in cubic yards only, and all embedded materials become incidental to the concrete item. Therefore, they must be included in the concrete unit price by the contractor.

- When the design cross section is uniform, as it is in asphalt pavement, it is quicker to measure the linear feet of pavement completed than the area of the pavement.

CHAPTER 23

PRICE ESTIMATE

23.1   General

An engineer's cost estimate consists of two steps. Step one, discussed in Chapter 22, is to estimate the quantities as depicted in the design. The second step of the process is to estimate the prices for the work items identified in the design. As discussed in Chapter 19, the prices to be estimated are either unit prices or lump-sum prices. The estimation of construction costs by an engineer is challenging for a variety of reasons:

- Most civil engineers are trained in the technical and design side of civil engineering, but not the construction side. Because he or she understands based on past experiences what direct labor costs (man-hours), indirect costs, or outside services (e.g., drilling, testing, special consultants) are necessary, an engineer can prepare confidently an estimated cost for professional services for an engineering study or design. Similarly, a contractor can prepare confidently an estimated cost of a construction because he or she understands—based on past experiences—what direct labor, equipment and material costs, subcontractors, and other overhead and administration costs are necessary. When an engineer prepares an estimate of a construction, is he or she crossing the line and discipline into an area where he or she has little direct experience and training?

- Unlike building costs in architectural design, the costs of heavy civil engineering construction are greatly dependent on the types of work (highways, dams, tunnels, pipelines, etc.), locations of project sites, meteorological factors, subsurface conditions, etc. How to account for the impact of these factors on productivity will determine the accuracy of the cost estimate.

- The construction cost of a given project will change with the local construction market, bidding climate, availability of qualified contractors, wage requirements, and—for multiyear contracts—time. Anticipating and accounting for these factors will sometimes determine whether a project can go into construction.

- The size of the project will have significant influence on the prices. Historical price data, when applied indiscriminately and without adjusting for project size and duration, may result in significant errors on the actual cost of construction.

Recognizing these challenges, this chapter presents the practice of estimating civil engineering construction costs by an engineer.

23.2   Cost Components

Regardless of how a contractor is being compensated to construct a civil engineering project, the costs to him or her to construct that project can be broken down into the following categories:

Direct labor cost: Direct labor includes all wages and salaries that are paid to the construction crew directly involved with the construction. The crew may include project manager, superintendents, foremen, crew chiefs, equipment operators, and laborers. Overtime costs to workers are also included in this category.

Equipment cost: Equipment cost includes the cost of owning, leasing, and operating the equipment that is used in construction. This cost is not to be confused with the permanent equipment cost, which is part of the completed project.

Products and materials costs: These costs include the cost to purchase the contractor-furnished products and materials required in the construction, including delivery charges and all handling and storage costs. Some products (such as special mechanical gates and valves) require manufacturer's representative to be on site during installation. That cost should also be included in the product cost.

Subcontractor cost: The subcontractor cost to a general contractor will include the cost of paying the subcontractors plus a markup for overhead and profit.

Mobilization and demobilization: Includes movement of personnel, equipment, and supplies to the project site; establishment of offices, buildings, plants, and other facilities at the project site; temporary utilities and temporary access roads; temporary site protection; permits and licenses; periodic and final cleanup; equipment not chargeable to a specific task; and demobilization of personnel and equipment.

Bonds, insurance, taxes: Bonds will include bid bond, performance bond, and payment bond. Insurance will include workers' compensation, liability, builder's risk, etc. Taxes will include payroll tax, sales tax, and other local, state, and federal taxes.

Overhead and profit: Overhead costs include home office expenses, communication, business insurance and taxes, management cost, finance cost, etc. The profit is the financial return of the contractor's investment, and provides the contractor with an incentive to perform the work as efficiently as possible.

All of the costs incurred by the contractor will be charged to the owner in one way or another. There are many ways these individual costs are paid to the contractor, depending on the preference of different owners and design engineers. The following is a description of one way that the contractor can be compensated for these costs:

- The cost of mobilization and demobilization is paid for as a lump-sum item or separate lump-sum items. A schedule of payment (see, e.g., Section 19.5) is usually set up to allow the contractor to recuperate these initial costs as quickly as possible to minimize the practice of unbalanced bidding (see Section 25.3).

- The contractor is rarely paid separately for direct labor, equipment, and products and materials unless it is under a *force account* arrangement. Rather, payment items are set up to compensate the contractor for the units of work completed (such as cubic yards of fill placed, or cubic yards of concrete placed), which include all labor, equipment, and materials, including subcontractor cost. When the contractor's work is difficult to quantify for payment, lump-sum items (such as dewatering, erosion control, and stream diversion) are set up and a partial payment will be made to the contractor based on an estimate of percentage of items completed.

- Bonding and insurance costs are a direct cost to the project, and should be returned to the contractor as quickly as possible. A lump-sum bid item for bonds and insurance can be set up to allow the contractor to recuperate these costs in the first progress payment.

- No separate payment items are usually allowed for taxes, overhead, and profit, so the contractor would need to spread these costs over other scheduled paid items.

Division 1 costs, the costs of mobilization and demobilization, vary from site to site and from project to project. In general, the larger the contract, the higher the mobilization and demobilization cost. For preliminary cost-estimating purposes, one can assume the mobilization and demobilization as a percentage of the total construction cost. Figure 23-1 shows the cost of mobilization and demobilization of over 55 bids on small to medium-sized dam projects, expressed as a percentage of the total construction bids. These data show that the mobilization and demobilization costs are in the range of 5% to 15% of the total cost. Even though these data are based on dam projects only, they can be applied to some of the similar

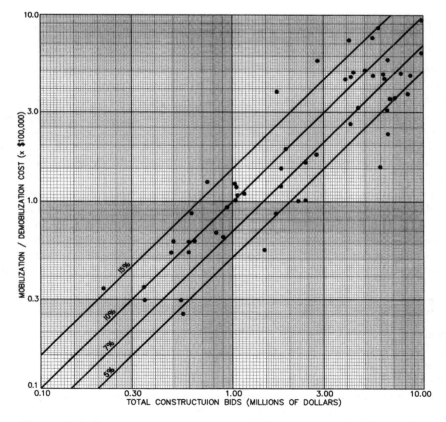

Figure 23-1.   Mobilization and demobilization costs for dam projects.

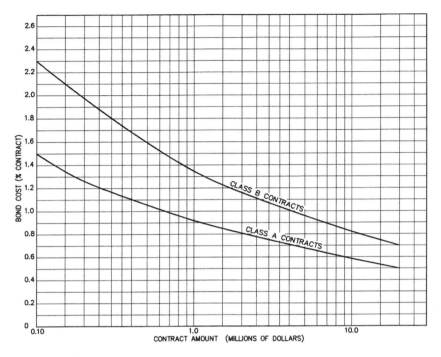

Figure 23-2.   Cost of bonding for heavy civil construction projects.

types of heavy civil construction, such as dikes, levees, stream bank protection, road work, and revetment work.

The premium for performance and payment bonds is dependent on the types of civil engineering construction work and contract amount. Figure 23-2 shows the average bonding cost as a percentage of the contract amount, based on 2001 Means *Heavy Construction Cost Data* (R.S. Means Co. 2001). Two cost curves are shown: one for Class A contracts and another for Class B contracts. In general, the larger the contract, the lower the percentage of bond cost. Class A contracts include construction of highways, roads, bridges, as well as street paving, airport runways, and river bank protection. Class B contracts include construction of all types of buildings, repair work on buildings, dams, dikes, levees, wharves, sewers, pipelines, pilings, subways, tunnels, and revetments. Figure 23-2 can be used to estimate the approximate bonding cost for cost-estimating purposes. The actual cost of the bonding to a particular contractor depends on the bond rating of that contractor, which varies from contractor to contractor. Different bonding companies also offer somewhat different rates.

23.3   Engineer's Approach

With few exceptions, the traditional method engineers use to estimate unit prices and lump-sum prices is based on data from past projects, published construction cost data, quotations from manufacturers, suppliers and specialty contractors, and judgment and experience. Perhaps the main characteristic of this approach is that cost components, as outlined in Section 23.2, are not identified separately. Rather, each of the estimated prices already includes direct costs (labor, equipment, and materials), indirect costs, overhead, and profit. In fact, in this approach, there is no attempt to break down these separate costs. When all of the cost items are added together for a particular project, it is assumed that all of the costs outlined in Section 23.2 will be accounted for.

Contrary to quantity takeoffs, price estimates should be performed by experienced personnel (e.g., an engineer or a cost-estimate professional). This person, who is referred to as the *cost estimator*, needs to have a thorough understanding of the design and construction requirements, including:

Site conditions: The characteristics of the site affects the contractor's staging and stockpile areas, temporary access requirements and constraints, availability of water and power, selection of types of equipment and construction processes, environmental limitations, labor housing and transport, and restrictions by local, state, and federal agencies.

Work requirements: Work requirements are based on construction drawings and technical specifications. However, understanding the work requirements and costing the work requires an understanding of construction practices, constructability, and construction schedules. For example, to estimate the cost of foundation dewatering, the cost estimator needs to know the following:

- Duration of dewatering

- Method of dewatering (sump-pumping, well points, deep wells)

- Layout of dewatering facilities

- Operating and maintenance costs

None of this information typically is contained in the construction documents, and would need to be evaluated as part of the cost-estimating process.

Another example is to borrow and process clean sand and gravel material to manufacture a filter material using on-site bank-run materials. Costs for this unit price item may include:

- Excavating and hauling bank-run materials to the processing plant

- Screening the bank-run materials to remove oversized materials

- Crushing the oversized material

- Washing the screened materials to remove excessive fines

- Stockpiling the processed materials

Understanding construction equipment and processes: Selecting the proper equipment and construction processes to meet specified requirements is a decision that the cost estimator must make to form a basis for the cost. Because there is usually more than one way to build something, the assumed method merely represents a reasonable scenario that should be practical, constructible, and price-competitive. The following are some examples:

- If a borrow site is so steep and small that only backhoes and loaders can be used, it is not reasonable to assume scraper production of the borrow material, even though scrapers would result in a lower unit price.

- If cast-in-place concrete is being placed in a steep terrain that is not accessible to transit-mix trucks, additional support equipment (e.g., cranes, conveyors, pumping equipment) will be required, which will increase the unit price of the concrete.

After the cost estimator identifies the work requirements, site limitations, equipment and construction method for a particular work item, the next step is to assign dollar values to represent each work component. To obtain these costs, the cost estimator needs the following information:

- Purchased material cost—Materials or products that are purchased should be based on quotations from the material suppliers or on recent cost data for similar materials. Material costs are usually quoted in free on board (FOB) price, which corresponds to the point at which the supplier will deliver goods without a delivery charge to the buyer. For manufactured goods, FOB usually means at the factory. For processed earthwork materials and aggregates, FOB usually means at the pit or quarry. Therefore, delivery costs would need to be added to the FOB cost to obtain the delivered price.

- Borrow material cost—Costs for materials that are available on site should include labor and equipment to excavate, process (if necessary), and haul the materials to the temporary stockpiles or to the placement locations.

- Installation cost—The installation cost includes labor and equipment to install or place the required work. Labor and equipment costs for Division 1 through Division 16 work are published annually by several sources, such as R.S. Means Company (see Section 23.4), Richardson, and Building News. Because published data do not normally provide adequate depths and scopes for specialized construction items—slurry walls, post-tensioned anchors, underwater diver work, foundation improvements, tunnels, etc.—it is advisable to consult specialty contractors. When it is assumed that a certain portion of the work will be performed by subcontractors, such as specialty contractors, a subcontractor markup should be added to the quotes given by these sources. For the engineer's cost-estimating purpose, a 5% to 10% markup can be used.

- Productivity of labor and equipment—The estimate of production rates or labor forces and equipment requires significant construction experience and familiarity with equipment performance. In heavy civil construction, the work is accomplished primarily with mechanized equipment, with the exception of concrete work, which involves significant labor effort in placing forms and reinforcements. Production rates of equipment depend on many factors, including ease or difficulty of access, size of the project, types of materials being handled, and working conditions, such as weather and time of year. Considerable experience and judgment are required to estimate production rates.

23.4 Means Cost Data

Construction cost data is published monthly and annually by many sources, including *Engineering News-Record*, R.S. Means Company, Inc., Richardson, Building News, and state transportation departments. The most comprehensive and most commonly used source for heavy civil construction projects, discussed in detail in this section, is published by R. S. Means Company, Inc.

R. S. Means Company, Inc. provides a comprehensive list of construction information, products, and services in North America and worldwide. Among the services are a variety of cost data books, such as *Building Construction Cost Data, Facilities Construction Cost Data, Heavy Construction Cost Data,* and *Labor Rates for Construction Cost Industry*. Many of these are also available in electronic medium. For heavy civil construction projects, the relevant reference is the *Heavy Construction Cost Data* (R.S. Means Co. 2001), which will be discussed in more details in this section.

The Means cost data in any given reference are organized in accordance with the Construction Specifications Institute (CSI) system (see Chapter 18). The use of CSI format in Means data provides a direct reference between design documents' technical specifications and costs. Costs are divided into 16 CSI divisions, and then are further broken down into Level Two and Level Three reference numbers. For example, the CSI reference number for site grading is 02310. That same reference number can be found in Means cost references for the same construction work.

For good documentation practice, it is important to properly cite the reference number for a particular cost item from the Means data base. Means uses the term *Line Number* as a reference. The line number is a 12-digit code based on the 5-digit CSI MasterFormat classification, as follows:

02315-300-6220

The first 5 digits are the CSI Level Three section number. The remaining digits are the Means designation. The line number used above refers to compaction by four passes of a vibratory roller in 6-inch lifts. The section number 02315 is used in CSI for excavation and fill.

For each line item, a great amount of information can be found in the Means cost data:

- Type of equipment and construction method

- Type of construction crew

- Production rate of crew and equipment

- Unit of work production

- Material, equipment, and labor costs, with and without overhead and profit

In selecting the appropriate cost item, it is important that the cost estimator understands what equipment type, crew type, and construction method are practical. This is why construction experience is important in cost estimating. Even though most engineers are not trained in the application, operation, and productivity of construction equipment, cost engineers should at least become familiar with the uses and limitations of commonly used construction equipment. Equally important is the use of a proper construction crew. In Means data, the type and size of the construction crew are defined for each cost item. Details of the crew include total number of labor categories (such as foreman, laborers, equipment operator), type and number of equipment (such as truck, loader, chain saws, tractor), hourly and daily production output, and total daily cost for each crew. This information is important in evaluating whether the crew assumed in the line item is reasonable for the work. In some situations, such as structure demolition and groundwater dewatering, it is difficult to estimate the cost of work in production units (such as cubic yards), and it is necessary to estimate the cost based on the estimated production rates of an appropriate crew. The following are two examples of construction crews used in Means:

Crew B-11M: 1 Equipment operator

                        1 Laborer

                        1 Backhoe loader (80 H.P.)

Crew C-14F: 1 Laborer foreman

                            2 Laborers

                            6 Cement finishers

                            1 Gas-engine vibrator

The Means cost data represent the national average based on 30 major U.S. cities. For a given type of construction work, the cost is different from city to city and from region to region. The estimated costs for a particular project need to be adjusted for this geographic factor. Means uses the term *City Cost Index* (CCI) to make this adjustment for 305 cities in the United States. The CCI is different for Division 2 through 16 work, and is also separated for materials and labor. The adjustment is made as follows:

Specific city cost = CCI × National average cost

As an example, the 2001 CCIs for a project located in San Antonio, Texas, are:

Division 2: Site construction   CCI = 0.896

Division 3: Concrete   CCI = 0.729

Division 15: Mechanical   CCI = 0.875

The following examples illustrate how the Means cost data are used to estimate construction cost components for a project located in San Antonio, Texas. The cost data used are based on 2001 Means *Heavy Construction Cost Data.*

1.   Open-cut earth excavation
   2001 Means, 02315-400-1300
        Front-end loader, 3 C.Y. capacity $1.30/C.Y.
   2001 Means, 02320-200-0310
        Dump truck, 12 C.Y. capacity, 1/4-mile round-trip $2.86/C.Y.
   Subtotal: $4.16/C.Y.
   Adjust for CCI, Division 2 work   ×0.896
   Adjusted unit price: $3.73/C.Y.
   Round off: $3.75/C.Y.

2.   Placing embankment earthfill, using on-site borrow
   2001 Means, 02315-400-1300
        Front-end loader, 3 C.Y. capacity $1.30/C.Y.
   2001 Means, 02320-200-0310
        Dump truck, 12 C.Y. capacity, 1/4-mile round-trip $2.86/C.Y.
   2001 Means, 02315-410-5020
        Dozer, 300 H.P., spread fill $0.99/C.Y.
   2001 Means, 02315-300-5720
        Sheepsfoot roller, 12″ lift, 4 passes $0.46/C.Y.
   Subtotal: $5.61/C.Y.
   Adjust for CCI, Division 2 work   ×0.896
   Adjusted unit price: $5.03/C.Y.
   Round off: $5.00/C.Y.

3. Furnishing and placing concrete grade walls
   2001 Means, 03310-240-4260

Concrete in place, including forms, rebars, finish		$265.00/C.Y.
Adjust for CCI, Division 3 work	×0.729	
Adjusted unit price:		$193.19/C.Y.
Round off:		$200.00/C.Y.

4. Furnishing and installing PVC slotted drain pipes
   2001 Means, 02530-780-2040

6″ PVC pipe in place, SDR 35, unslotted		$5.45/L.F.
Cut slots to meet specifications (per quote)		$2.50/L.F.
Subtotal:		$7.95/L.F.
Adjust for CCI, Division 2 work	×0.896	
Adjusted unit price:		$7.12/L.F.
Round off:		$7.25/L.F.

23.5   Other Considerations

Estimating the cost of construction does not stop at quantity takeoff and price estimate. Numerous other factors should be considered that might affect construction costs:

Wage Requirements

The labor costs contained in most published cost data, such as the Means cost data, are typically based on average prevailing wages for construction trades. If a project mandates that Davis-Bacon wages be used, such as in federal construction, adjustment would need to be made for this wage difference. Also, published labor information does not include overtime costs. If the schedule of a project is such that significant overtime is anticipated, then the overtime costs should be factored into the labor costs.

Availability of Contractors

Before a project goes into bid, it is advantageous for the owner to evaluate qualified local, regional, and national contractors' level of general interest in his or her project. In general, the more interest the project gets from contractors, the more favorable will be the bids. The reverse is also true. There are two separate issues regarding the availability of qualified contractors:

- Number of qualified contractors available if the project requires special construction skills and experience.

- Availability of the contractors during the bidding period. This bidding climate is a supply-and-demand issue. When contractors are busy, there is less tendency for them to submit competitive bids, and the bids will generally come in higher. The reverse is also true.

Based on the results of the owner's research, the engineer's cost estimate may need to be adjusted up or down to account for these factors.

Site Factors

In civil engineering construction, each site is unique and poses a set of factors that will affect the cost of the construction. Some of these factors will be accounted for in the mobilization and demobilization costs, and others will affect individual work items, such as stream diversion or dewatering. The following site factors tend to increase the cost of construction:

- Adverse weather conditions, such as frequent precipitation, and short construction season.

- Short contract duration, which limits the contractor's flexibility on manpower and equipment, and would result in higher labor cost from overtime.

- Remote site, which limits the availability of laborers. A short local labor supply will increase labor costs because the contractor will need to bring in workers from other areas. A remote site also increases the haul costs of materials and supplies.

- Difficult access, which requires significant improvement to temporary access roads or new roads.

- Traffic problems and extensive traffic control in an urban setting or downtown setting.

- Off-site disposal requirements, which increase haul and disposal costs in landfills.

- Quarry or borrow-pit permit requirements. These contractor permits are generally costly, and approval for these permits is time-consuming because of necessary compliance with numerous environmental laws.

- Excavation adjacent to high groundwater, such as adjacent to a stream, river, or lake.

- Restriction on how a site can be developed. The limits on an area that the contractor can develop at any given time may affect the types of equipment that can be used, as well as the production rates of construction crews.

- Restriction on working hours. Contractors prefer to work from sunrise to sunset, and on weekends when necessary. Limits on working hours may increase the number of workdays if the contractor cannot take advantage of all of the daylight for construction.

- Restriction on noise level while working. This restriction may limit the contractor on the type of equipment that can be used. For example, if diesel pumps cannot be used for dewatering, then the more expensive electric pumps would be needed.

Size of Project

The unit rates and production rates obtained from published cost data, such as the Means cost data, represent average productivity and may even represent somewhat idealized working conditions. These conditions may be achievable for medium to large projects, but may not be possible for small projects. For small projects, the unit rates from published data or other historical data should be adjusted up, sometimes by a significant amount, to account for small quantities and low production rates. For example, an estimated unit rate of $3.00 per cubic yard for open-cut excavation may be reasonable when the quantity is 10,000 to 20,000 cubic yards, but—because the same equipment is mobilized to perform significantly less work—may be too low for 300 cubic yards. When quantities are low, the published costs can be misleading, and should be used with caution. In some cases, the unit rate for a small quantity may be several hundred percent higher than that estimated for a large quantity in published databases.

23.6   Cost-Risk Analysis

For large projects, cost estimators sometimes perform a cost-risk analysis to identify and measure the cost impact of project uncertainties on the estimated construction cost. An important step from this process is the application of the so-called 80-20 Rule. According to the 80-20 Rule, approximately 80% of the cost of a project is contained in approximately 20% of the estimated work items. The author has applied this rule to his own database, and concurs that this rule is generally accurate. This rule allows the cost estimator to focus his or her effort on about 20% of the work items that control most of the project cost. The cost estimators of even small to medium projects are encouraged to apply this rule to evaluate the estimated construction costs while considering the factors discussed in Section 23.5.

CHAPTER 24

ALLOWANCES AND CONTINGENCIES

24.1 General

This chapter discusses various contingency factors relating to design and construction projects and cost allowances that are normally used to account for these contingencies. Contingencies represent degrees of uncertainty that result from a variety of causes. Some are inherent and unavoidable, and some can be minimized. There are two types of contingencies: *design contingency* and *construction contingency*. Allowances for contingencies are carried from the conception of a project through conceptual and final design and into construction. As a project progresses from the planning level through design, allowances for contingencies also change. In general, the higher the degree of uncertainty in a project, the higher will be the contingency allowance. Also discussed in this chapter is the allowance for future cost escalation, which should be included in budgets for projects for which construction will not begin immediately after the effective date of the construction cost estimate.

24.2 Design Contingency

Design contingency is associated with the following uncertainties:

- Uncertainty in the definition of the project, especially at the early stages of project design. Design contingency can cover small changes in the scope of the project, but is not intended for significant changes. When a project scope or definition changes significantly, the engineer is responsible for advising the owner that the project cost should be updated for the new scope.

- Uncertainty from a lack of design details, which is typical at early levels of design. The absence of detailed design information may be a result of incomplete site characterization—such as an inadequate topographic base map or inadequate subsurface exploration—or it may be a conscious effort to explore several alternatives that are feasible; and in-depth details may not be developed when alternatives are screened for technical feasibility, constructibility, and cost.

- Changes in design concept or design details. The design contingency should only be used to accommodate small changes in design concept and design details. When major design changes are anticipated, it is advisable to update the cost estimate for the new concept.

- Construction cost items that are not yet identified (*unlisted items*). In feasibility design or conceptual design, only the major cost items (see Section 23.6 for the 80-20 Rule) are identified, and minor cost items are accounted for in the design contingency as unlisted items.

- Approximate nature of quantities of identified cost items because of incomplete details available for a more accurate quantity takeoff. It is not unusual to have the quantities based on only a typical section for each design alternative.

In general, allowances for design contingency should decrease as design information becomes better defined. In other words, design contingency should be largest during a planning level or reconnaissance level design, and should be zero at the completion of final design. Different engineers or owners have different guidelines on allowances for design contingency. The following are the recommended guidelines:

Level of Design	*Contingency Allowance*
Planning level, reconnaissance level	20% to 30%
Conceptual design level	15% to 25%
Final design, 30% complete	5% to 10%
Final design, 75% complete	0% to 5%
Final design, 100% complete	0%

The contingency factors recommended above can be changed somewhat on a case-by-case basis, depending on the uncertainties that are unique to each project. It is important that no design contingency is left at the end of final design, because that would imply that the design is incomplete. When a particular design feature is expected to change during construction because of differing site conditions, the cost increase should be accounted for in the construction contingency discussed in Section 24.3.

24.3 Construction Contingency

The construction contingency allowance in the engineer's cost estimate is associated with the following uncertainties:

- Uncertainty related to local construction market, bidding climate, and availability of interested contractors during bidding and construction period. This uncertainty almost always exists each time a project is being bid. In some cases, an unfavorable bidding climate may result in either very few bids or very high bids, to a point at which the owner cannot afford the project. That is why it is important to research this uncertainty during design (see Chapter 23) and to adjust for these market factors in the engineer's cost estimate. The construction contingency can only absorb small adjustments in market factors. When a major portion of the construction contingency is spent to account for the market factors, it leaves little allowance for the construction. This is not advisable.

- Uncertainty related to differing site conditions—softer foundation, deeper or shallower bedrock surface, etc. Differing site conditions may directly affect the plan quantities (e.g., common excavation becomes rock excavation), and they also may result in design changes with cost implications. The design engineer can minimize this uncertainty during design by an adequate site characterization investigation. The cost and effort spent for an adequate site characterization during design more than offsets the cost increase during construction caused by differing site conditions.

- Uncertainties caused by unforeseen conditions, such as unusually wet or cold construction weather, flooding, or other "acts of God" that cannot be predicted, and should not be normally included in a project construction cost estimate. Risk associated with unforeseen conditions should be born by the owner and not by the contractor. Experience has shown that insistence of an owner to transfer all the risk cost for unforeseen conditions to the contractor only results in claims, litigation, and ruined working relationships. In most cases, court decisions favor the contractor. In addition, a contractor is likely to submit a higher bid if he or she has to take all risks.

- Changes associated with inaccuracies in quantities or errors in plan quantities. As design engineer, this uncertainty can be minimized by a proper cost-estimating effort during design (see Chapters 22 and 23).

- Errors in design that would result in changes in construction method and delay that would justify change orders and associated cost increase. The design engineer can minimize this uncertainty with a thorough quality control and review of the construction documents prior to bid (see Sections 9.10 and 15.4).

Regardless of the quality of construction documents, a project should never go into construction without some allowance for construction contingency. Despite competent design effort and careful review, construction drawings and specifications are never perfect. Plan quantities are estimated with accuracies of up to a few percent, but most importantly, subsurface conditions and other site conditions for heavy civil construction projects will always contain some level of uncertainty, regardless of how conscientiously the designer tries to characterize the project site.

For civil engineering projects, it is recommended that a 5% to 15% contingency factor be included in the construction budget. The actual contingency depends on the uncertainties and risk to be assessed by the engineer for each particular project. An average of 10% is used for most projects.

24.4   Escalation Adjustment

Escalation is the increase in project cost to account for future inflation. There are three situations in which the project cost should be adjusted for escalation:

1. A project that was designed a few years ago is ready for construction. In this case, the project cost estimated in the past would need to be escalated to the current construction year.

2. A project has been designed now and is expected to go into construction some years in the future. In such a case, the project cost estimated now would need to be escalated to the future construction year.

3. An ongoing project will be constructed over many years.

There are many published sources from which inflation factors can be obtained. For civil engineering construction purpose, the commonly used source in the profession is the cost indices published by the *Engineering News-Record* (ENR). The ENR publishes a Construction Cost Index (CCI) and a Building Cost Index (BCI) every month. These indices are based on average costs of labor and materials for 20 cities in the United States and 2 cities in Canada. Only the CCI is discussed here. The CCI has been compiled by ENR since 1908, with a base index of 100 in Year 1913.

The following is used to compute the escalation factor from Year A to Year B:

Escalation Factor = (CCI for Year B)/(CCI for Year A)

Two examples are given below to illustrate the estimation of escalation factors for the two situations cited above.

Example 24.1—Estimate current project cost for a design completed in 1990 that will be under construction in 2000.

CCI for 1990 = 4732

CCI for 2000 = 6222

Escalation Factor = 6222/4732 = 1.31

Therefore, the project cost estimated in 1990 should be increased by 31%.

Example 24.2—Estimate future project cost for a design completed now (say 2004), that will be under construction in 2008.

The ENR only provides 12-month projections. The escalation factor will need to be estimated. Based on records since 1980, the average U.S. annual inflation is approximately 3% to 4%. Assume 3% annual inflation:

Escalation Factor = 1.03^4 = 1.13

Therefore, the project cost currently estimated should be increased by 13% for future construction.

CHAPTER 25

EVALUATION OF BIDS

25.1   General

Traditionally, the design phase of a project ends with the completion of construction draw-ings, technical specifications, and the engineer's cost estimate. Bid solicitation is tradition-ally viewed as part of the procurement phase of project development and it occurs after the plans and specifications are completed. As discussed below, the design phase sometimes does not end during bid solicitation. When a project schedule is tight, some owners start the bid solicitation process—advertisement, prebid conference and site visit, bid amendments, questions and answers—before the design is finalized. The design is finalized during bid so-licitation, and changes to the original construction documents are made in bid amend-ments. For a variety of reasons, this practice should be discouraged, but it is commonly used by owners to buy additional time for design. Significant revisions to the plans and specifi-cations during bidding should be avoided for the following reasons:

- Major changes to the design confuse the bidders and can result in misunderstand-ings of work requirements and erroneous bids.

- Major changes to the design may complicate issues during construction, as both the contractor and construction manager would need to go back to bid amendments instead of the original documents. The use of conformed drawings (see Sections 4.2 and 13.1) eliminates this problem.

- Major changes during bidding always result in a prolonged bidding process and sometimes may not even have any time-saving advantages.

In any case, after bids are received and opened, they are usually summarized and eval-uated by the owner, frequently with the assistance of the design engineer. The evaluation and analysis of bids may reveal errors in the bids, unbalanced bidding, or items that re-quire bid verification to obtain additional information. When the owner is satisfied with the bids, a contractor is selected based on price, qualifications, or both. This chapter dis-cusses the process of summarizing, evaluating, and verifying bids. The topic of unbalanced bidding is introduced, as this practice is sometimes used by contractors for a variety of strategic reasons.

25.2   Bid Summary

Bids for civil engineering construction projects are usually summarized in a *bid tab*, which is a tabulation of all bid prices identified in the bid schedule and submitted by each of the bidders. In general, bid tabs contain the following information:

- Name of the project and bid opening date

- All the information in the bid schedule, including bid item number, item description, units, and plan quantities

- The engineer's cost estimate, representing the owner's estimate of the project cost

- The names of the bidders and their respective bid prices and total costs

For publicly funded projects, bid tabs are considered public information and may be available from the public agency upon request. For privately funded projects, bid tabs may be considered confidential information and may not be available to the public. In any case, data from bid tabs are generally evaluated and analyzed by the owner and his or her engineer before a bidder is selected. Bid data is valuable information to a cost engineer or cost estimator for future projects and represents a valuable resource.

Table 25-1 is an example of a bid tab for a fictitious project for illustration purposes. The project is for rehabilitation of an existing stream channel. The work consists of 14 bid items, including dewatering and stream diversion, excavation, backfill, reinforced concrete hydraulic control structures, channel lining, and reclamation. Three bids are received, and their bids, along with the owner's estimate, are summarized in the table. The engineer's cost estimate is \$244,175, which is within the bid range of \$231,953 to \$308,693. The apparent low bidder is Bidder No. 2, with a bid of \$231,953.

25.3   Unbalanced Bidding

To maximize profit during construction, contractors sometimes use *unbalanced bidding* to establish a contractual advantage on prices during bidding. A bid with unbalanced bidding still maintains the competitiveness as far as the total bid amount is concerned, because the bidder still needs to win the project to execute his or her strategy. Unbalanced bidding is not illegal, and a bid with unbalanced bidding cannot be thrown out for that reason only. The practice, however, borders on unethical and is generally viewed negatively by owners and engineers.

In unbalanced bidding, the prices of certain bid items are distorted so that they do not reflect the actual costs to the contractor. It is used to gain the following advantages during construction:

- *Cash-flow issue*—The prices of work items that will be performed first are artificially inflated so that the contractor can be paid for these work items early, and thus improve cash-flow management and finance of the construction.

- *Quantity difference*—When a bidder recognizes that the quantity is significantly underestimated on a particular work item, that bidder may deliberately over-inflate the bid price to obtain a large profit on that item. In other words, the bidder is taking advantage of an engineer's estimating error.

Table 25-1. Little Clear Creek rehabilitation bid tabulation

Item	Item description	Quantity	Unit	Engineer's estimate		Bidder no. 1		Bidder no. 2		Bidder no. 3	
				Unit price	Amount	Unit price	Amount	Unit price	Amount	Unit price	Amount
1	Mobilizing/Demobilization	1	Lump Sum	$10,000	$10,000	$18,000	$18,000	$12,000	$12,000	$13,000	$13,000
2	Dewatering and Stream Diversion	1	Lump Sum	$20,000	$20,000	$5,000	$5,000	$25,000	$25,000	$18,500	$18,500
3	Clearing and Grubbing	1	Lump Sum	$10,000	$10,000	$8,800	$8,800	$7,500	$7,500	$6,800	$6,800
4	Sediment and Erosion Control	1	Lump Sum	$3,000	$3,000	$2,500	$2,500	$1,500	$1,500	$4,400	$4,400
5	Unclassified Excavation	1500	Cubic Yard	$4.50	$6,750	$9.00	$13,500	$5.00	$7,500	$4.25	$6,375
6	Rock Excavation	100	Cubic Yard	$25.00	$2,500	$75.00	$7,500	$22.00	$2,200	$30.00	$3,000
7	General Fill	1,850	Cubic Yard	$5.50	$10,175	$6.75	$12,488	$5.15	$9,528	$3.55	$6,568
8	Structural Fill	500	Cubic Yard	$7.50	$3,750	$10.25	$5,125	$8.85	$4,425	$6.25	$3,125
9	Reinforced Concrete	300	Cubic Yard	$300.00	$90,000	$350.00	$105,000	$275.00	$82,500	$375.00	$112,500
10	Riprap	1500	Cubic Yard	$40.00	$60,000	$65.00	$97,500	$35.00	$52,500	$45.00	$67,500
11	Riprap Bedding	500	Cubic Yard	$25.00	$12,500	$45.00	$22,500	$22.50	$11,250	$27.75	$13,875
12	New Asphalt Walkway	1000	Square Feet	$3.00	$3,000	$3.25	$3,250	$2.75	$2,750	$4.00	$4,000
13	Placing Topsoil	400	Cubic Yard	$25.00	$10,000	$10.70	$4,280	$24.50	$9,800	$28.50	$11,400
14	Seeding and Mulching	1	Acre	$2,500	$2,500	$3,250	$3,250	$3,500	$3,500	$1,875	$1,875
	Total (Item 1–14)				$244,175		$308,693		$231,953		$272,918

Note: bid opening date: October 17, 1987

The bid tabulation in Table 25-1 is used to illustrate the concept of unbalanced bidding. For this example, Bidder No. 1 is employing the tactic of unbalanced bidding in his or her bid. Specifically, the following bid items are unbalanced:

Item 1—Mobilization/Demobilization	$18,000/LS
Item 2—Dewatering and Stream Diversion	$5000/LS
Item 5—Unclassified Excavation	$12.00/CY
Item 6—Rock Excavation	$75.00/CY
Item 13—Placing Topsoil	$10.70/CY

The strategy for Bidder No. 1 is as follows:

- Unbalance the costs of Items 1 and 2 to improve cash flow, so that the combined cost of these two items is equal to the actual costs. For this contract, 90% of the cost for mobilization and demobilization will be paid early, whereas the cost for dewatering and diversion will be distributed throughout the project.

- Unbalance the costs of Items 5 and 13 to improve cash flow, so that the combined cost of these two items is equal to the actual costs. Earth excavation will need to be performed early, and placing topsoil for reclamation will be performed at the end of the project.

- Unbalance the unit price of Item 6 because the bidder recognizes that the plan quantity for rock excavation will be exceeded by at least 50%.

Table 25-2 shows how the costs are being shifted between Items 1 and 2, and between Items 5 and 13.

As illustrated in Table 25-2, the exercise of shifting costs from one item to another does not change the actual costs to the contractor for performing the work, and thus maintains his or her competitive edge. However, his or her strategy to over-inflate the unit price of rock excavation is an added risk to the bidder, because the amount for that item is now over-inflated and becomes less competitive with the other bidders. Nevertheless, because the quantity is low, the bidder decides that this risk, typical for contractors, is worth taking.

When the bids are compared among one another and with the engineer's cost estimate, the unbalanced bidding of Bidder No. 1 becomes obvious. His or her prices are either significantly higher or significantly lower than the other prices. In this particular example, Bid-

Table 25-2. Example of unbalanced bidding

	Item	Actual cost	Unbalanced cost	Difference
1.	Mobilization/Demobilization	1 LS @ $8,000	1 LS @ $18,000	+$10,000
2.	Dewatering/Stream Diversion	1 LS @ $15,000	1 LS @ $5000	−$10,000
5.	Unclassified Excavation	1500 CY @ $3.85/CY = $5575	1500 CY @ $9.00/CY = $13,500	+$7725
13.	Placing Topsoil	400 CY @ $30.00/CY = $12,000	400 CY @ 10.70/CY = $4,280	−$7720
			Net Difference:	+$5.00

der No. 1 is not the low bidder, and the owner does not have to deal with the implications of the unbalanced bidding. If the bid for Bidder No. 1 were the apparent low bid, Bidder No. 1 may be selected as the contractor, and the owner would need to readjust his or her construction cash flow to deal with this paid schedule. As for the inflated unit price for rock excavation, the engineer should reestimate the quantity for rock excavation, and a design change may be required to minimize the increase in construction cost caused by a much higher rock excavation.

25.4 Bid Verification

Bid verification is the process of evaluating a bid by requesting additional information on bid items that are significantly different from the engineer's cost estimate. When a bid price is significantly different from the engineer's estimate, one of the following scenarios will likely occur:

- The bidder misunderstands the scope of work.

- The engineer has made an error in his or her quantity or price.

- Unbalanced bidding has occurred.

- The bidder understands the scope of work, but his or her method of construction is different, his or her material supplier provides a different quote, and the production rate is different.

- The bidder deliberately under-priced or over-priced that item because of the market condition, and his or her busy schedule.

The main purpose of bid verification is to make sure that the bid reflects the contract scope of work, and that no error has been made unintentionally. All of the other reasons listed above, including unbalanced bidding, are not grounds to disqualify a bid. Of course, if an engineer discovers that he or she has made a mistake in estimating the price or quantity of a particular item, that matter is between the owner and the engineer.

Assuming that the contractor is being selected based on price only, the owner is interested in verifying the apparent low bid first. Typically, one or more of the following pieces of information will be requested from the apparent low bidder on bid items to be verified:

- Schedule of values (breakdown of materials, equipment, and labor cost, especially for lump-sum bid items)

- Work to be produced by those bid items (types of materials, locations, limits, criteria, etc.)

- Sources and quotations of materials and products

- Schedule of work for those bid items

- Quantities that are estimated independently by the bidder

Only when the owner is satisfied with the information provided by the apparent low bidder, and no errors are discovered, will the apparent low bid become the low bid, and the contract will be awarded to the low bidder. The construction phase begins when the contract is awarded or, sometimes, when notice to proceed is given to the contractor.

REFERENCES

American Society of Civil Engineers (ASCE). 1979. "Summary report of questionnaire on specifications (owner and owner representative returns)." *Journal of the Construction Division, Proceedings of the American Society of Civil Engineers* 105(CO3), pp. 163–186.

American Society of Civil Engineers (ASCE). 1999. *Topographic surveying.* Reston, Va.: ASCE Press.

American Society of Civil Engineers (ASCE). 2002. "Do not neglect site information furnished by the owner." *Civil Engineering* 104(9).

Anderson, J. M., and Mikhail, E. M. 1997. *Surveying, theory and practice,* 7th Ed. New York: McGraw-Hill.

Brinker, R. C. 1969. *Elementary surveying.* Scranton, Penn.: International Textbook Company.

Church, H. K. 1981. *Excavation handbook.* New York: McGraw-Hill.

Clayton, C. R. I., Simons, N. E., and Matthews, M. C. 1982. *Site investigations.* New York: Halsted Press.

Colorado Department of Transportation (CDOT). 1999. *Standard Specifications for Road and Bridge Construction.*

Construction Specifications Institute (CSI). 1996a. "Construction documents fundamentals and formats module, FF/020: Formats." Manual of Practice, Alexandria, Va.

Construction Specifications Institute (CSI). 1996b. "Construction documents fundamentals and formats module, FF/070: Specifications." Manual of Practice, Alexandria, Va.

Construction Specifications Institute (CSI). 1996c. "Construction documents fundamentals and formats module, FF/100: General requirements." Manual of Practice, Alexandria, Va.

Construction Specifications Institute (CSI). 1996d. "Construction documents fundamentals and formats module, FF/090: Coordinating drawings and specifications." Manual of Practice, Alexandria, Va.

Construction Specifications Institute (CSI). 1996e. "Construction documents fundamentals and formats module, FF/130: Product evaluation." Manual of Practice, Alexandria, Va.

Construction Specifications Institute (CSI). 1996f. "Construction documents fundamentals and format module, FF/170: Specification language." Manual of Practice, Alexandria, Va.

Construction Specifications Institute (CSI). 1970. "System for relating drawings and specifications (Document MP-1D)." Manual of Practice, Alexandria, Va.

Construction Specifications Institute (CSI). 1995. *MasterFormat: Master list of numbers and titles for the construction industry.* Alexandria, Va.

Construction Specifications Institute (CSI). 1997. *SectionFormat: A recommended format for construction specifications sections.* Alexandria, Va.

Construction Specifications Institute (CSI). 1999. *PageFormat: Recommended page formats for construction specifications.* Alexandria, Va.

Dixon, S. A., ed. 1998. "Lessons in professional liability." *DPIC's loss prevention handbook for design professionals.* Monterey, Calif.: DPIC Companies.

Engineering News-Record (ENR). 2000. "No Stamp of Approval on Building Plans," pp. 34–46.

211

Engineers Joint Contract Documents Committee (EJCDC). 1990. "Standard general conditions of the construction contract." *EJCDC Document No. 1910-8.*

Fisk, E. R. 1992. *Construction project administration,* 4th Ed. Englewood Cliffs, N.J.: Prentice Hall.

Gambatese, J. A. 2000. "Safety in a designer's hands." *Civil Engineering.* 70(6), pp. 56–59.

Giesecke, F. E., Mitchell, A., Spencer, H. C., Hill, I. L., and Loving, R. O. 1975. *Engineering graphics,* 2nd Ed. New York: Macmillan.

Hansen, K. D., and Reinhardt, W. G. 1991. *Roller compacted concrete.* New York: McGraw-Hill.

Head, K. H. 1980. *Manual of soil laboratory testing. Volume 1: Soil classification and compaction tests. Volume 2: Permeability, shear strength and compressibility tests.* London: Pentech Press Limited.

Hvorslev, M. J. 1949. "Subsurface exploration and samplings for civil engineering purposes." *Report for Committee on Sampling and Testing, Soil Mechanics and Foundations Division, American Society of Civil Engineers.* Vicksburg, Miss.: Waterways Experiment Station.

Legget, R. F., and Karrow, P. F. 1983. *Handbook of geology in civil engineering.* New York: McGraw-Hill.

Occupational Safety and Health Administration (OSHA). 1990. "Construction safety and health regulations." *Code of Federal Regulations Title 29, Part 1926.*

Ringwald, R. C. 1993. *Means heavy construction handbook.* Kingston, Mass.: R.S. Means Company, Inc.

R. S. Means Company, Inc. 2001. *Heavy construction cost data,* 15th Ed., Annual Edition. Kingston, Mass.: R. S. Means Company, Inc.

Rosen, H. J. 1999. *Construction specifications: writing, principles and procedures,* 4th Ed. New York: Wiley.

Spangler, M. G., and Handy, R. L. 1982. *Soil engineering.* New York: Harper & Row.

State of Arizona. 1991. *Code and rules of the state board of technical registration for architects, assayers, engineers, geologists, landscape architects, and land surveyors.* December 18.

State of North Dakota. 2000. *Registration news.* Vol. 3.2 North Dakota State Board of Registration for Professional Engineers and Land Surveyors.

State of Texas. 1991. *Law and rules concerning the practice of engineering and professional engineering registration.* Austin, Tex.: Texas State Board of Registration for Professional Engineers. September 1.

Terzaghi, K., Peck, R. B., and Mesri, G. 1996. *Soil mechanics in engineering practice.* New York: Wiley.

Winterkorn, H. F., and Fang, H.-Y. 1975. *Foundation engineering handbook.* New York: Van Nostrand Reinhold Co.

EXAMPLE SPECIFICATIONS FOR REFERENCE DATA PRESENTATION

SECTION 17110
SUBSURFACE DATA

PART 1—GENERAL:

1.1 SECTION INCLUDES

A. Detailed boring logs prepared by BasePoint Design Corporation for borings performed in March 1999. Locations of the borings are shown on the Drawings. Simplified boring logs are also shown on the Drawings.

B. Well installation logs for five open-standpipe observation wells. Locations of the wells are shown on the Drawings.

C. Groundwater monitor data in the observation wells.

D. Detailed test pit logs in Borrow Area A and Borrow Area B. The locations of the test pits are shown on the Drawings. Simplified logs of the test pits are also shown on the Drawings.

1.2 LIMITATIONS

A. The Owner does not represent that the logs and other data show the conditions that will be encountered in performing the Work. The Owner represents only that such information shows conditions at the locations of the borings and test pits. It is understood that the making of the deductions, interpretations, and conclusions from all the accessible factual information, including the nature of the materials to be excavated, the difficulties of making and maintaining the required excavations, and the difficulties of doing other work affected by the geology and other subsurface conditions at the site of the work, are the Contractor's sole responsibility.

PART 2—PRODUCTS:

Not used.

PART 3—EXECUTION:

Not used.

SECTION 17210
SLURRY WALL MIX TEST DATA

PART 1—GENERAL:

1.1 SECTION INCLUDES

A. Laboratory test report by ABC Corporation of reservoir water for suitability as mix water for biopolymer slurry for drain wall construction.

B. Laboratory test report by ABC Corporation on soil-bentonite mixes for soil-bentonite slurry wall construction.

1.2 RELATED SECTIONS

A. Section 02262:   Drain walls.

B. Section 02265:   Soil-bentonite walls.

PART 2—PRODUCTS:

Not used.

PART 3—EXECUTION:

Not used.

SECTION 17410
CLIMATOLOGICAL DATA

PART 1—GENERAL:

1.1 DESCRIPTION

A. This section contains climatological data for a weather station located in the vicinity of the project site. The weather data were obtained from the National Oceanic and Atmospheric Administration, National Climatic Data Center.

B. The weather station is located approximately 2 miles southeast of the project site at an elevation of 1345 feet above sea level. This section contains climatic data for 1977 through 1997. The following climatic data is contained in this section.

1. Mean daily minimum and maximum temperatures for each month.

2. Mean monthly temperature.

3. Highest and lowest daily temperatures for each month.

4. Mean number of days for each month in which the maximum temperature was 90°F or above.

5. Mean number of days for each month in which the maximum temperature was 32°F or below.

6. Mean number of days for each month in which the minimum temperature was 32°F or below.

7. Mean number of days for each month in which the minimum temperature was 0°F or below.

8. Mean monthly precipitation.

9. Greatest monthly precipitation.

10. Greatest daily precipitation for each month.

11. Mean and maximum monthly snowfall for each month.

PART 2—PRODUCTS:

Not used.

PART 3—EXECUTION:

Not used.

GLOSSARY

Azimuth: Angles measured clockwise from any meridian, ranging from 0 degrees to 360 degrees.

Bank Run: An informal term used in heavy civil construction and aggregate industry to denote natural state or composition of materials obtained directly from the ground without any processing. Also referred to as *Pit Run*.

Base Bid: A list of bid items in the bid schedule that the owner is obligated to award in a construction contract.

Benchmark: A permanent reference point with known elevation and datum, and may also include horizontal coordinates.

Bid Amendment: A modification to bidding documents, issued during bidding.

Bid Additive: An optional paid item in a bid schedule in additional to the base bid.

Bid Alternative: An alternative work item in a bid schedule to replace one or more items in the base bid. Also referred to as *Bid Option*.

Bid Schedule: Breakdown of construction work items, accompanied, where appropriate, by estimated quantities that are used to solicit prices from contractors and later used as a basis for contractor compensation during construction.

Bid Tab: A compilation or summary of bid prices from all bidders, usually including the engineer's cost estimate.

Bond: A sum of money paid in advance to guarantee a certain obligation, such as performance of a construction contract or payment of bills for labor and materials.

Catch Line: Intersection of an excavated surface or fill slope with the ground surface.

Catch Point: Intersection of an excavated line or fill line with the ground surface.

Change Order: Documentation of a change of scope of work during construction; may or may not accompany a change in contract price or contract schedule.

Claim: A request from a contractor for compensation on work not scheduled for payment in the contract.

Clay Shale: Sedimentary rock composed mainly of silt- and clay-sized particles, typically weakly cemented and with medium to high plasticity.

Clearing: Removal of trees, brush, grass, debris, rocks, or other surface materials; typically done prior to excavation in the ground.

Conformed Drawings: Construction drawings incorporating all design changes made during bidding; issued for construction after award of bid.

Constructibility: Ability for a construction method to be built safely and using conventional materials and equipment.

Contour: Line of equal elevation.

Davis-Bacon Wages: Wages mandated through the Davis-Bacon Act of 1931, which is a federal labor law requiring that mechanics and laborers receive not less than prevailing wages as determined by the Secretary of Labor.

Declination: Horizontal angle between the magnetic meridian and the true or geographic meridian.

Descriptive Geometry: Science of graphic representation and the solution of spatial relationships of points, lines, and planes by means of projections.

Dewatering: Removal of groundwater through pumping, sumps, ditches, well points, deep wells, or other methods.

Easting: North-south grid lines in a horizontal coordinate system, with coordinates increasing in the easterly direction.

Engineering Geology: Discipline of applying knowledge of geology to engineering problems.

Force Account: A method of construction procurement (also known as *Day-Labor Method*) in which the contractor is compensated for actual labor and equipment through hourly rates.

Freeboard: A vertical distance between the waterline and the top of the structure such as a canal bank or a dam crest.

FOB (Free on Board): A point of delivery of materials and supplies without delivery charge to the buyer.

Geomembrane: Any of the many manufactured synthetic fabrics, typically with low permeability, used in geoenvironmental and geotechnical applications.

Glacial Till: Sediments deposited by a glacier, typically unsorted and unstratified.

Grade: A term used in heavy civil construction to denote a ground surface.

Ground Penetration Radar: A technique used in geophysical field investigations to locate underground pipes, voids, structural defects, concrete structures, and other subsurface features.

Grubbing: Removal of the bottom portion of vegetation, such as tree stumps and roots.

Invert: A control point or surface in a structure, generally at the bottom, such as the inside bottom of a pipe or culvert.

Lacustrine Deposits: Sediments deposited in a lake environment, typically fine-grained silts and clays.

Layer: A term used in computer-aided drafting (CAD) to denote a group of information on a drawing, such as controls, vegetation, utilities, etc.

Legend: An explanatory caption on a construction drawing, usually for symbols, lines, or hatched patterns.

Mensuration: The measurement of geometric quantities.

Micron: A dimension equal to one-millionth of a meter.

Monument: A surface reference point used for survey control. A monument can be a *Benchmark*.

Neat Line: A term used in heavy civil construction and design to denote the design limit of a feature, such as limit of excavation, fill, and concrete.

Northing: East-west grid lines in a horizontal coordinate system, with coordinates increasing in the northerly direction.

Offset: A distance measured perpendicular to a baseline or centerline.

Outlet Works: An appurtenant structure in a dam with the function of controlling water releases through the dam.

Pit Run: See *Bank Run*.

Photogrammetry: The science of obtaining measurements and other qualitative information from photographs.

Plan Holder: A person, firm, or other entity on a recorded list, who has an interest during bidding of a construction project and is issued a set of bid documents and all amendments.

Poly-line: A term used in computer-aided drafting (CAD) to denote a continuous, three-dimensional line consisting of straight-line segments with known coordinates.

Plan Quantity: Estimated quantity of a construction paid item based on the design shown on the construction drawings; used as the basis during bidding.

Planimeter: An instrument used for estimating areas on a drawing by manually tracing the perimeter of the area.

Pro-Bono: Voluntary work by a professional without a service fee.

Record Drawings: Construction drawings issued after construction that incorporate all design changes made during construction.

Redline: A document with marked-up review comments or changes.

Riparian Vegetation: Vegetation that grows adjacent to a stream or lake.

Rock Quality Designation (RQD): The RQD is used in rock mechanics to quantify the in-situ rock quality, and is defined as the sum of the rock cores 4 inches in length or longer, expressed as a percentage of the total core run.

Roller-Compacted Concrete (RCC): A lean, zero-slump concrete placed using earth-moving equipment in lifts. The aggregate for the concrete typically contains a significant amount of gravel and relatively low fines contents.

Running Average: An average of two or more consecutive numbers, starting with the most current number.

Schedule: A tabulation of information on a construction drawing or specifications.

Screening: A technique of using lighter line weights or light objects as background on a design drawing.

Seismotectonic Assessment: Engineering geology study intended to evaluate faults and sources and magnitudes of seismic loads.

Shop Drawings: Construction drawings prepared by a contractor during construction, showing installation details and other construction details and descriptions not shown in the design that are within the responsibility and expertise of the contractor.

Soil Cement: A lean, zero-slump mixture of soil and cement placed using earth-moving equipment in lifts. The aggregate for the soil cement typically contains a relatively low gravel content and high fines content.

Spillway: An appurtenant structure in a dam used to pass flood flows through the dam.

Standard Dimension Ratio (SDR): Ratio of the outside diameter of a pipe to its wall thickness.

Standard Penetration Test Blowcounts: A field test procedure in foundation engineering in which the number of blows of a standard-size hammer is used to drive a standard-size soil sampler into the ground; generally used to evaluate in-situ density or consistency of the materials.

Subgrade: A surface in the ground, generally obtained through excavation, used for structural support.

Unconfined Compressive Strength: Axial strength of a soil or rock specimen by axial loading without lateral confinement.

Unwatering: Removal of surface water by means of sumps, ditches, pumping, drainpipes, or other methods.

Value Engineering: A study or evaluation of alternatives that satisfy the original function and need at the lowest life-cycle cost.

LIST OF RESOURCES

Resource Name	Address	Telephone	Fax	Website
American Society of Civil Engineers (ASCE)	1801 Alexander Bell Drive, Reston, VA 20191	(800) 548-2723	(703) 295-6222	www.asce.org
American Council of Engineering Companies (ACEC)	1015 15th Street N.W., 8/F Washington, DC 20005	(202) 347-7474	(202) 898-0068	www.acec.org
Associated General Contractors of America (AGC)	333 John Carlyle Street, Suite 200 Alexandria, VA 22314	(703) 534-3118	(703) 538-3119	www.agc.org
American Water Works Association (AWWA)	6666 W. Quincy Avenue Denver, CO 80235	(303) 794-7711	(303) 794-3951	www.awwa.org
American Society for Testing and Materials (ASTM)	100 Barr Harbour Drive P.O. Box C700 West Conshohocken, PA 19428	(610) 832-9585	(610) 832-9555	www.astm.org
American Concrete Institute (ACI)	38800 Country Club Drive Farmington Hills, MI 48331	(248) 848-3700	(248) 848-3701	www.aci-int.org
American National Standards Institute (ANSI)	1819 L Street N.W., 6/F Washington, DC 20036	(202) 293-8020	(202) 293-9287	www.ansi.org
American Institute of Steel Construction (AISC)	One East Wacker Drive, Suite 3100 Chicago, IL 60601	(312) 670-2400	(312) 670-5403	www.aisc.org
American Association of State Highway & Transportation Officials (AASHTO)	444 North Capitol Street N.W., Suite 249 Washington, DC 20001	(202) 624-5800	(202) 624-5806	www.transportation.org
BNi Building News	1612 S. Clementine Street Anaheim, CA 92802	(888) 264-2665	—	www.bnibooks.com
Construction Specifications Institute (CSI)	99 Canal Center Plaza, Suite 300 Alexandria, VA 22314	(800) 698-2900	(703) 684-0465	www.csinet.org
Construction Specifications Canada (CSC)	120 Carlton Street, Suite 312 Toronto, Ontario M5A 4K2 Canada	(800) 668-5684	(416) 777-2198	www.csc-dcc.ca
Engineering News-Record (ENR)	Two Penn Plaza, 9/F New York, NY 10121	(888) 867-8208	(212) 904-3150	www.enr.com
National Society of Professional Engineers (NSPE)	1420 King Street Alexandria, VA 22314	(703) 684-2800	(703) 836-4875	www.nspe.org
National Climatic Data Center	151 Patton Avenue Asheville, NC 28801	(828) 271-4800	(828) 271-4876	www.ncdc.noaa.gov
Post-Tensioning Institute	8601 N. Black Canyon Hwy., Suite 103 Phoenix, AZ 85021	(602) 870-7540	(602) 870-7541	www.post-tensioning.org
Portland Cement Association (PCA)	5420 Old Orchard Road Skokie, IL 60077	(847) 966-6200	(847) 966-8389	www.portcement.org
R.S. Means Co., Inc.	63 Smiths Lane P.O. Box 800 Kingston, MA 02364	(800) 334-3509	(800) 632-6732	www.rsmeans.com
Richardson Engineering c/o Aspen Technology	Ten Canal Park Cambridge, MA 02141	(617) 949-1000	(617) 949-1030	www.aspentech.com

INDEX